D0855353

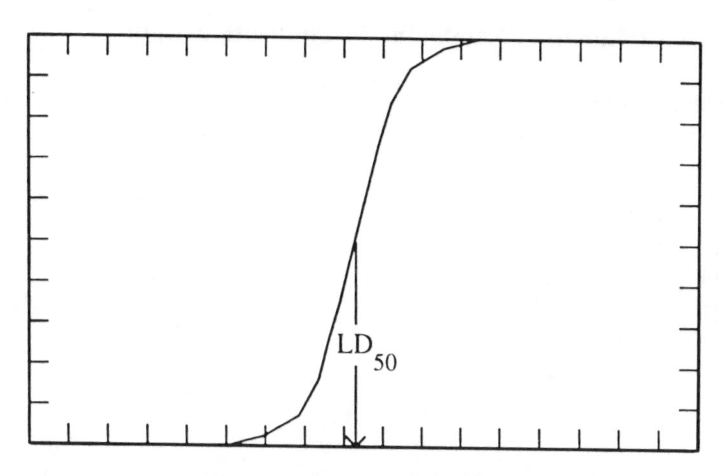

TOXICOLOGICAL CHEMISTRY

A GUIDE TO TOXIC SUBSTANCES IN CHEMISTRY

STANLEY E. MANAHAN

Library of Congress Cataloging-in-Publication Data

Manahan, Stanley E.
Toxicological chemistry

Bibliography: p.
Includes index.
1. Toxicological chemistry I. Title.
RA1219.3.M36 1988 615.9'00154 88-25552
ISBN 0-87371-149-1

LEWIS PUBLISHERS, INC.
121 South Main Street, Chelsea, Michigan 48118

PRINTED IN THE UNITED STATES OF AMERICA

Stanley E. Manahan is a Professor of Chemistry at the University of Missouri – Columbia, where he has been on the faculty since 1965. He received his A.B. in chemistry from Emporia State University in 1960 and his Ph.D. in analytical chemistry from the University of Kansas in 1965. Since 1968 his primary research and professional activities have been in environmental chemistry and have included development of methods for the chemical analysis of pollutant species, environmental aspects of coal conversion processes, development of coal products useful for pollutant control, hazardous waste treatment, and toxicological chemistry. He teaches courses on environmental chemistry, hazardous wastes, and toxicological chemistry and has lectured on these topics throughout the United States as an American Chemical Society Local Section tour speaker.

Professor Manahan has written books on environmental chemistry, applied chemistry and quantitative chemical analysis. He has been the author or co-author of approximately 70 research articles.

CONTENTS

PREFACE

This book was written to bridge the gap between toxicology and chemistry. It defines toxicological chemistry as the science which deals with the chemical nature and reactions of toxic substances; their origins and uses; and the chemical aspects of their exposure, transformation, and elimination by biological systems. Although toxicological chemistry must consider all major areas of toxicology, it emphasizes the chemical formulas, structures, and reactions of toxic substances.

This book has developed from the author's teaching and research in the area of environmental chemistry. The ultimate concern in environmental chemistry is the effect of pollutant chemicals on living systems, a topic addressed by toxicological chemistry. The material in the book was assembled as part of the author's research, consulting, and activities as an expert witness, primarily with hazardous wastes. This material was used in preliminary form as the text for a course at the University of Missouri – Columbia entitled "Topics in Environmental and Toxicological Chemistry" and taken by graduates and undergraduates in a variety of disciplines.

Toxicological Chemistry is designed for use by people in a variety of disciplines and with a wide range of backgrounds. It provides brief introductions to toxicology, chemistry, organic chemistry, and biochemistry, so that readers without substantial expertise in one or more of these areas can still understand the remainder of the book. To retain a chemical emphasis, it is organized primarily on the basis of classes of inorganic and organic chemical substances. Among those who should find the book useful are toxicologists (for its chemical content), chemists (for its toxicological aspects), engineers, industrial hygienists, research laboratory personnel dealing with hazardous and toxic substances, people working with the disposal and elimination of hazardous substances, regulatory personnel, and any others who, for various reasons, need information about the relationship between chemicals and their toxic effects.

The author would appreciate hearing from readers regarding corrections to the book, uses of the book, and suggestions for future editions. It is hoped that this work will fill an important niche at the interface between toxicology and chemistry.

ACKNOWLEDGMENTS

The author is grateful to Dean Milton Glick (now Provost at Iowa State University) for providing a Macintosh SE computer at an early date to enable word processing and preparation of structures and figures for this book. He would also like to acknowledge the assistance of personnel from Softshell International who freely and patiently gave information and backup for the Chemintosh structure drawing program with which the structures and illustrations in this book were drawn. Over a several year period students in the author's "Environmental Chemistry" and "Topics in Environmental and Toxicological Chemistry" courses have provided valuable input and suggestions for this work. Robin Berry and the rest of the staff of Lewis Publishers have been outstanding to work with in preparing this book. A special thanks is due to Anne Manahan, who worked long hours assisting in the production of the final manuscript.

1

Toxicology and Toxicological Chemistry

1.1. POISONS AND TOXICOLOGY

A **poison**, or **toxicant**, is a substance that is harmful to living organisms because of its detrimental effects on tissues, organs, or biological processes. **Toxicology** is the science of poisons. These definitions are subject to a number of qualifications. Whether or not a substance is poisonous depends upon the type of organism exposed, the amount of the substance, and the route of exposure. In the case of human exposure, the degree of harm done by a poison can depend strongly upon whether the exposure is to the skin, by inhalation, or through ingestion. For example, a few parts per million of copper in drinking water can be tolerated by humans. However, at that level it is deadly to algae in their aquatic environment, whereas at a concentration of a few parts per *billion* copper is a required nutrient for the growth of algae. Subtle differences like this occur with a number of different kinds of substances.

History of Toxicology

The origins of modern toxicology can be traced to M. J. B. Orfila (1787-1853), a Spaniard born on the island of Minorca. In 1815 Orfila published a classic book,[1] the first ever devoted to the harmful effects of chemicals on organisms. This work discussed many aspects of toxicology recognized as valid today. Included are the relationships between the demonstrated presence of a chemical in the body and observed symptoms of poisoning, mechanisms by which chemicals are eliminated from the body, and treatment of poisoning with antidotes.

Since Orfila's time, the science of toxicology has developed at an increasing pace with advances in the basic biological, chemical, and biochemical sciences. Prominent among these advances are modern instruments and techniques for chemical analysis that provide the means for measuring chemical poisons and their metabolites at very low levels and with remarkable sensitivity, thereby greatly extending the capabilities of modern toxicology.

The toxic effects of chemicals have gained increasing recognition in legislation designed to protect the public from chemical hazards, such as the Toxic Substances Control Act.[2] For example, under the Emergency Planning and Community Right-to-Know provision of the 1986 Superfund Amendments and Reauthorization Act, concerns using or processing substantial amounts of specified chemicals and members of chemical groups were required to provide estimates of toxic chemicals released to the environment by July 1, 1988.[3] Compilation of this Toxics Release Inventory is a massive effort involving approximately 30,000 facilities in the U. S.

1.2. CLASSIFICATIONS OF TOXICOLOGY

Toxicology can be classified in several ways. One such system distinguishes the areas of environmental, economic, and forensic toxicology. Each of these divisions is summarized briefly below.

Environmental Toxicology

Environmental toxicology deals with exposure to toxic substances through polluted air or water, contaminated food, or industrial materials. Ozone in an atmosphere in which photochemical smog[4] has formed or carbon monoxide in air heavily contaminated by automobile exhaust gases are examples of substances to which organisms are involuntarily exposed in the atmosphere. Other examples of toxicants in the environment are arsenic compounds in drinking water, pesticide residues in food, and nitrosamines in industrial cutting oils. Environmental toxicology involves the sources, conditions, effects, and safety limits of incidental exposure of organisms to toxic contaminants in air, water, or materials that are handled.

Economic Toxicology

Economic toxicology has to do with the harmful effects of substances administered intentionally to organisms or biologic tissue for a beneficial effect. Perhaps the most common example of economic toxicology is the study of undesirable side effects of drugs (therapeutic agents). Some of the ways in which a therapeutic agent may be toxic include harmful effects on non-target tissue or a non-target organ, allergic reactions, and acting as the agent was designed to act, but to an excessive degree. An important aspect of economic toxicology is that in which a chemical is used to eliminate an **uneconomic species** to the benefit of an **economic species**. Thus, for example, selectively toxic herbicides are applied to soil in corn fields to kill uneconomic weeds, thereby allowing the economic corn to grow free of competition for space, light, nutrients, and water.

Forensic Toxicology

The medical and legal aspects of toxicology are classified as **forensic toxicology**. This category covers both intentional and accidental exposure to toxic substances. The medical branch of forensic toxicology includes the diagnosis and treatment of the effects of toxic substances. Specific disease conditions caused from toxic substance exposure are within the realm of **clinical toxicology**. A major part of forensic toxicology deals with treatment for exposure to toxic substances, including treatment procedures and **antidotal agents** that counteract the effects of harmful chemicals. A burgeoning part of the legal branch of forensic toxicology is that of legal liability because of harmful effects from accidental exposure to toxicants, most often in the workplace.

Other Classification Systems

Industrial toxicology is the branch of toxicology that deals with workplace exposure to toxicants.[5,6,7] In a broader sense industrial toxicology involves aspects of forensic, environmental, and economic toxicology.

Scientifically it is reasonable to classify toxicology according to the parts of the body affected or by toxic effect. The research components of the U.S. National Toxicology Program are so categorized.[8] **Cellular toxicology** is the study of ways in which toxicants alter cells in potentially detrimental ways. Closely related is **genetic toxicity**, which explores alterations of DNA by toxicants. **Carcinogenesis** investigations have as their ultimate objective establishment of the potential for toxicants to cause cancer. **Reproductive and developmental toxicology** is the study of the effects of chemicals on the reproductive system and the developing embryo. As other examples, **renal toxicology** pertains to effects of toxicants on the kidney and **pulmonary toxicology** applies to the lungs and respiratory system. **Immunotoxicology** is growing in importance with increasing recognition of the effects of toxicants on the immune system.

1.3. TOXICOLOGICAL CHEMISTRY

As the term implies, **toxicological chemistry** is the chemistry of toxic substances with emphasis upon their interactions with biologic tissue and living organisms. Toxicological chemistry deals with the chemical nature and reactions of toxic substances and involves their origins, uses, and chemical aspects of exposure, fates, and disposal.[9] By necessity, toxicological chemistry must consider non-chemical areas of toxicology. However, the emphasis in this discipline remains upon the chemical formulas and reactions of toxic substances. Biochemical transformations of substances in living organisms are considered with emphasis upon the chemical natures of the reactants and products and how these chemical properties influence biotransformations. **Chemical disposition** of toxicants and classes of toxicants refers to their absorption, distribution, metabolism, and excretion and is closely related to toxicological chemistry. Figure 1.1 summarizes the relationship of toxicological chemistry to toxicology.

It has long been recognized that there is a strong relationship between the chemical properties of substances and their toxicities. The influence of the structures of molecules upon biological activity is especially important in determining this relationship. The term applied to the interface between chemical structure and biological activity is **structure-activity relationship, SAR.**[10] Because of the complexity and expense of performing biological tests, truly effective means to relate chemical properties to toxicity have the potential to make a significant contribution to toxicological science.[11] Especially with increasing sophistication of computerized computations, SAR calculations promise to become a very significant part of toxicological chemistry.[12]

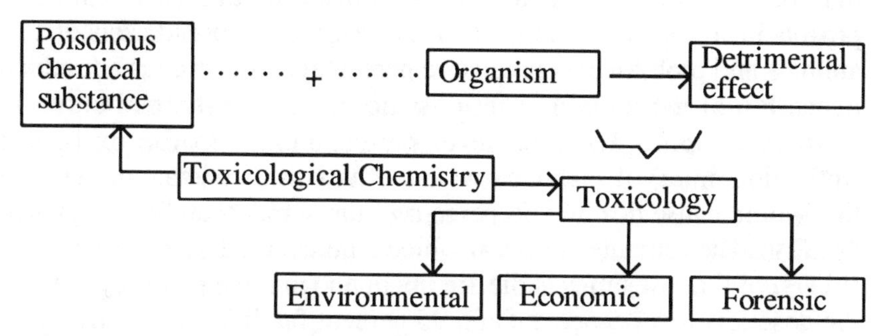

Figure 1.1. Toxicology as the science of poisons and the relationship of toxicological chemistry to toxicology.

1.4. DOSE-RESPONSE RELATIONSHIPS

Toxicants have widely varying effects upon organisms. Quantitatively, these variations include minimum levels at which the onset of an effect is observed, the sensitivity of the organism to small increments of toxicant, and levels at which the ultimate effect (particularly death) occurs in most exposed organisms. Some essential substances, such as nutrient minerals, have optimum ranges above and below which detrimental effects are observed.

Factors such as those just outlined are taken into account by the **dose-response** relationship, which is one of the key concepts of toxicology. **Dose** is the amount, usually per unit body mass, of a toxicant to which an organism is exposed. **Response** is the effect upon an organism resulting from exposure to a toxicant. In order to define a dose-response relationship, it is necessary to specify a particular response, such as death of the organism, as well as the conditions under which the response is obtained, such as the length of time from administration of the dose. Consider a specific response for a population of the same kinds of organisms. At relatively low doses, none of the organisms exhibits the response (for example, all live) whereas at higher doses, all of the organisms exhibit the response (for example, all die). In between, there is a range of doses over which some of the organisms respond in the specified manner and others do not, thereby defining a dose-response curve.[13] Dose-response relationships differ among different kinds and strains of organisms, types of tissues, and populations of cells.

Figure 1.2 shows a generalized dose-response curve. Such a plot may be obtained, for example, by administering different doses of a poison in a uniform manner to a homogeneous population of test animals and plotting the cumulative percentage of deaths as a function of the log of the dose. The result is normally an **S**-shaped curve as shown in Figure 1.2. The dose corresponding to the mid-point (inflection point) of such a curve is the statistical estimate of the dose that would cause death in 50 percent of the subjects and is designated as LD_{50}. The estimated doses at which 5 percent (LD_5) and 95 percent (LD_{95}) of the test subjects die are obtained from the graph by reading the dose levels for 5 percent and 95 percent fatalities, respectively. A relatively small difference between LD_5 and LD_{95} is reflected by a steeper **S**-shaped curve and vice versa.

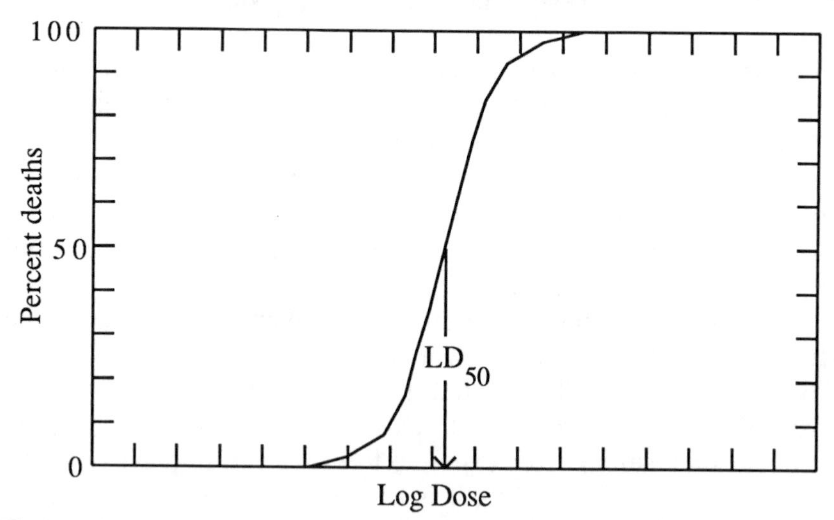

Figure 1.2. Illustration of a dose-response curve in which the response is the death of the organism. The cumulative percentage of deaths of organisms is plotted on the Y axis

Statistically, 68 percent of all values on a dose-response curve fall within ± 1 standard deviation of the mean at LD_{50} and encompass the range from LD_{16} to LD_{84}. Although not used in this book, the term **probit** is sometimes encountered in the literature; a probit of 5 stands for LD_{50} and differences of 1 probit unit correspond to 1 standard deviation.

1.5. RELATIVE TOXICITIES

Table 1.1 illustrates standard **toxicity ratings** that are used to describe estimated toxicities of various substances to humans. Reference is made to them in this book to denote toxicities of substances. Their values range from 1 (practically nontoxic) to 6 (supertoxic). In terms of fatal doses to an adult human of average size, a "taste" of a supertoxic substances (just a few drops or less) is fatal. A teaspoonful of a very toxic substance could have the same effect. However, as much as a quart of a slightly toxic substance might be required to kill an adult human.

When there is a substantial difference between LD_{50} values of two different substances, the one with the lower value is said to be the more **potent**. Such a comparison must assume that the dose-response curves for the two substances being compared have similar slopes (see Figure 1.2). If this is not the case, the substance for which the dose-response curve has the lesser slope may be toxic at a low dose where the other substance is not toxic at all. Put another way, the relative LD_5s of the substances may be reversed from the relative LD_{50}s.

Nonlethal Effects

So far, toxicities have been described primarily in terms of deaths of organisms, or lethality, an irreversible consequence of exposure. In many cases, **sublethal** and **reversible** effects are of greater importance. This is obviously true of drugs. By their very nature, drugs alter biologic processes; therefore, the potential for harm is almost always present. The major consideration in establishing drug dose is to find a dose that has an adequate therapeutic effect without undesirable side effects. A dose-response curve can be established for a drug that progresses from noneffective levels through effective, harmful, and even lethal levels. A low slope for this curve indicates a wide range of effective dose and a wide **margin of safety.** This term applies to other substances, such as pesticides, for which it is desirable to have a large difference between the dose that kills an uneconomic species and that which harms an economic species.

Table 1.1. Toxicity Scale with Example Substances.*

Substance	Approximate LD_{50}	Toxicity rating
	-10^5	1. Practically nontoxic, $>1.5 \times 10^4$ mg/kg
DEHP[1] →	–	
Ethanol →	-10^4	2. Slightly toxic, 5×10^3 – 1.5×10^4 mg/kg
Sodium chloride →	–	
Malathion →	-10^3	3. Moderately toxic, 500 – 5000 mg/kg
Chlordane →	–	
Heptachlor →	-10^2	4. Very toxic, 50 – 500 mg/kg
	–	
Parathion →	– 10	5. Extremely toxic, 5 – 50 mg/kg
	–	
TEPP[2] →	– 1	
	–	
Tetrodotoxin[3] →	-10^{-1}	
	–	
	-10^{-2}	6. Supertoxic, <5 mg/kg
	–	
TCDD[4] →	-10^{-3}	
	–	
	-10^{-4}	
	–	
Botulinus toxin →	-10^{-5}	

---

[1] Bis(2-ethylhexyl)phthalate; [2] Tetraethylpyrophosphate; [3] toxin from pufferfish; [4] TCDD represents 2,3,7,8,-tetrachlorodibenzodioxin, commonly called "dioxin."

* Doses are in units of mg of toxicant per kg of body mass. Toxicity ratings on the right are given as numbers ranging from 1 (practically nontoxic) through 6 (supertoxic) along with estimated lethal oral doses for humans in mg/kg. Estimated LD_{50} values for substances on the left have been measured in test animals, usually rats, and apply to oral doses.

1.6. REVERSIBILITY AND SENSITIVITY

Sublethal doses of most toxic substances are eventually eliminated from an organism's system. If there is no lasting effect from the exposure, it is said to be **reversible**. However, if the effect is permanent, it is termed **irreversible**. Irreversible effects of exposure remain after the toxic substance is eliminated from the organism. Figure 1.3 illustrates these two kinds of effects. For various chemicals and different subjects, toxic effects may range from the totally reversible to the totally irreversible.

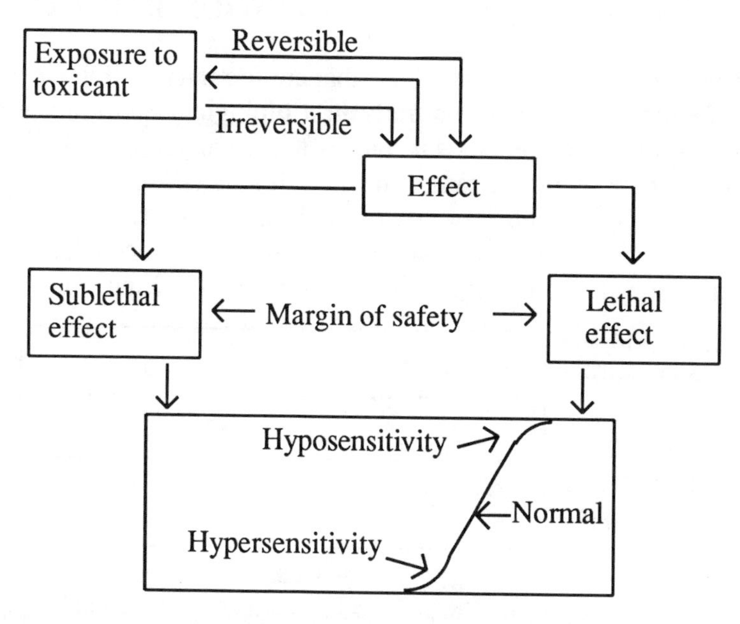

Figure 1.3. Effects of and responses to toxic substances.

Hypersensitivity and Hyposensitivity

Examination of the dose-response curve shown in Figure 1.2 reveals that some subjects are very sensitive to a particular poison (for example, those killed at a dose corresponding to LD_5), whereas others are very resistant to the same substance (for example, those surviving a dose corresponding to LD_{95}). These two kinds of responses define **hypersensitivity** and **hyposensitivity**, respectively;

subjects in the mid-range of the dose-response curve are termed **normals**. These variations in response tend to complicate toxicology in that there is not a specific dose guaranteed to yield a particular response, even in a homogeneous population.

In some cases hypersensitivity is induced. After one or more doses of a chemical, a subject may develop an extreme reaction to it. This occurs with penicillin, for example, in cases where people develop such a severe allergic response to the antibiotic that exposure results in death if countermeasures are not taken.

1.7. XENOBIOTIC AND ENDOGENOUS SUBSTANCES

Xenobiotic substances are those that are foreign to a living system, whereas those that occur naturally in a biologic system are termed **endogenous**. Endogenous substances are usually required within a particular concentration range in order for metabolic processes to occur normally. Levels below a normal range may result in a toxic response or even death, and the same effects may occur above the normal range. This kind of response is illustrated in Figure 1.4.

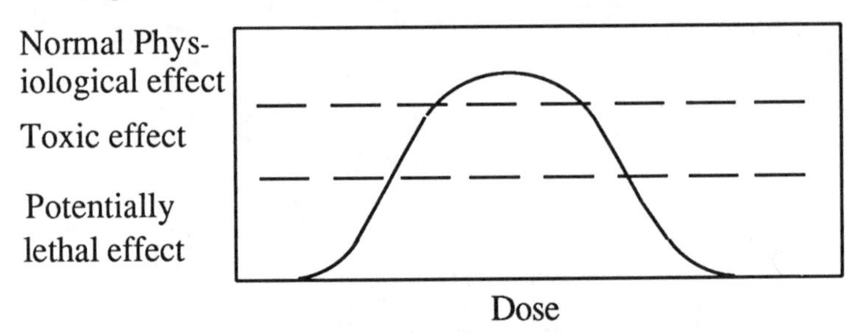

Figure 1.4. Biologic effect of an endogenous substance in an organism showing optimum level, deficiency, and excess.

Examples of Endogenous Substances

Examples of endogenous substances in organisms include various hormones, glucose (blood sugar), and some essential metal ions, including Ca^{2+}, K^{+}, and Na^{+}. The optimum level of calcium in human blood serum occurs over a rather narrow range of 9 – 9.5 milligrams per deciliter (mg/dL). Below these values a toxic response known as hypoglycemia occurs, manifested by muscle cramping. At serum levels above about 10.5 mg/dL hypercalcemia occurs, the major effect of which is kidney malfunction.

1.8. TOXICANT EXPOSURE

Before discussing toxicology further, it is useful to have in mind a general picture of the routes of exposure, effects, and elimination of toxicants by the body. A schematic of these pathways is shown in Figure 1.5. The overall scheme of exposure, body metabolism, and elimination of toxicants is so important that it composes Chapters 3 and 4 of this book. For the present, however, the outline shown in Figure 1.5 can serve as a basis of reference for the discussion of toxic substances.

1.9. KINETIC AND NONKINETIC TOXICOLOGY

Nonkinetic toxicology[14] deals with generalized harmful effects of chemicals that occur at an exposure site; a typical example is the destruction of skin tissue by contact with concentrated nitric acid, HNO_3. Nonkinetic toxicology applies to those poisons that are not metabolized or transported in the body or subject to elimination processes that remove them from the body. The severity of a nonkinetic insult depends upon both the characteristic of the chemical and the exposure site. Injury increases with increasing area and duration of the exposure, with the concentration of the toxicant in its matrix (for example, the concentration of HNO_3 in solution) and with the susceptibility of the exposure site to damage. The toxic action of the substance ceases when its chemical reaction with tissue is complete or when it is removed from the exposure site. Nonkinetic toxicology is also called **nonmetabolic** or **nonpharmacologic** toxicology.

Kinetic Toxicology

Kinetic toxicology, also known as **metabolic** or **pharmacologic** toxicology, involves toxicants that are transported and metabolized in the body. Such substances are called **systemic poisons** and they are studied under the discipline of **systemic toxicology**. Systemic poisons may cross cell membranes (see Chapter 2) and act upon **receptors** such as cell membranes, bodies in the cells, and specific enzyme systems. The effect is dose-responsive and it is terminated by processes that may include metabolic conversion of the toxicant to a metabolic product, chemical binding, storage (such as in fat or bone as shown in Figure 1.5), and excretion from the organism.

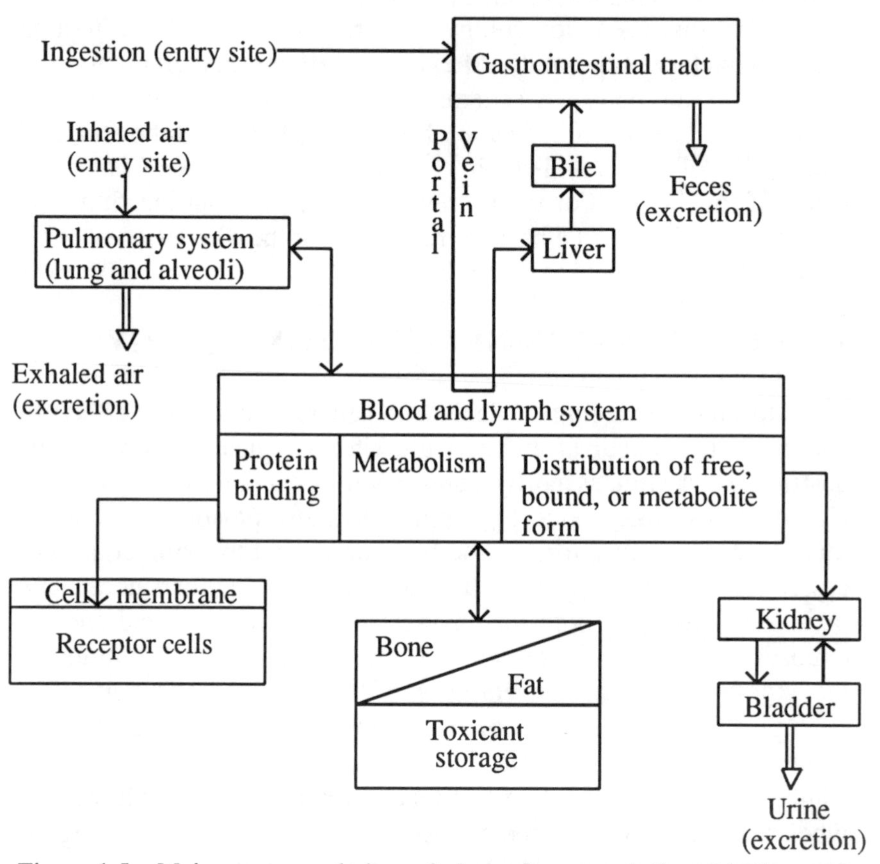

Figure 1.5. Major routes and sites of absorption, metabolism, binding, and excretion of toxic substances in the body.

In an animal a xenobiotic substance may be bound reversibly to a plasma protein in an inactivated form. A polar xenobiotic substance, or a polar metabolic product, may be excreted from the body in solution in urine. Nonpolar substances delivered to the intestinal tract in bile are eliminated with feces. Volatile nonpolar substances such as carbon monoxide tend to leave the body via the pulmonary system. As summarized in Figure 1.5, the ingestion, biotransformation, action on receptor sites, and excretion of a toxic substance may involve complex interactions of biochemical and physiological parameters. The study of these parameters within a framework of metabolism and kinetics is called **toxicometrics.**

1.10. RECEPTORS AND TOXIC SUBSTANCES

A toxic substance that enters the body through any of the entry sites shown in Figure 1.5 may undergo biochemical transformations that can increase or decrease its toxicity, affect its ability to traverse cell membranes, or enable its elimination from the body. A substance involved in a kinetic toxicological process (see Section 1.9) generally enters the blood and lymph system before it has any effect. Plasma proteins may inactivate the toxic substance by binding reversibly to it. The substance often undergoes biotransformation, which most commonly occurs in the liver, but may also take place in other types of tissue as well. These reactions are catalyzed by enzymes, most frequently mixed function oxidases. Toxicants can either stimulate or inhibit enzyme action. It is obvious that biochemical actions and transformations of toxicants are varied and complex. They are discussed in greater detail in Chapter 4.

Receptors

As noted in the preceding section, there are various *receptors* upon which xenobiotic substances or their metabolites act. In order to bind to a receptor, the substance has to have the proper structure or, more precisely, the right **stereochemical molecular configuration** (see Chapters 2 and 4). Receptors are almost always proteinaceous materials, normally enzymes. Nonenzyme receptors include opiate (nerve) receptors, gonads, or the uterus.

An example of a toxicant acting upon a receptor will be cited here; the topic is discussed in greater detail in Chapter 4. One of the most commonly cited examples of an enzyme receptor that is adversely affected by toxicants is that of **acetylcholinesterase**. It acts upon **acetylcholine** as shown by the reaction

$$\underset{\text{Acetylcholine}}{(CH_3)_3\overset{+}{N}-\underset{H}{\overset{H}{C}}-\underset{H}{\overset{H}{C}}-O-\overset{O}{\overset{\|}{C}}-\underset{H}{\overset{H}{C}}-H} + H_2O \xrightarrow{\text{Acetylcholinesterase}} (CH_3)_3\overset{+}{N}-\underset{H}{\overset{H}{C}}-OH + H-\underset{H}{\overset{H}{C}}-\overset{O}{\overset{\|}{C}}-OH \quad (1.1)$$

Acetylcholine is classified as a neurotransmitter. As such, it is a key substance involved with transmission of nerve impulses in the brain, skeletal muscles, and other areas where nerve impulses occur. An essential step in the proper function of any nerve impulse is its cessation (see Figure 1.6), which requires hydrolysis of acetylcholine by Reaction 1.1. Some xenobiotics, such as organophosphate compounds (see Chapter 13) and carbamates (see Chapter 10) inhibit acetylcholinesterase, with the result that acetylcholine accumulates and nerves are overstimulated. Adverse effects may occur in the central nervous system, autonomous nervous system, and at neuromuscular junctions. Convulsions, paralysis, and finally death may result.

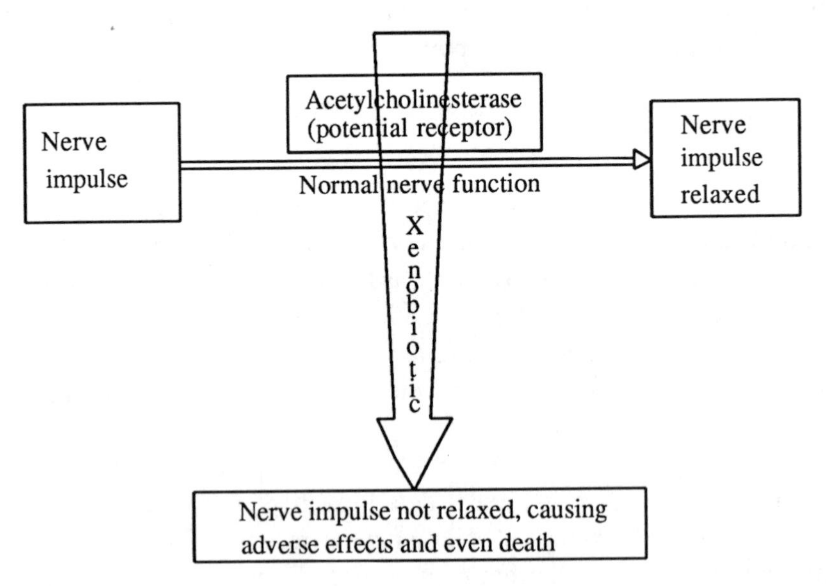

Figure 1.6. Example of a toxicant acting on a receptor to cause an adverse effect.

1.11. TOXICITY-INFLUENCING FACTORS

Classification of Factors

It is useful to categorize the factors that influence toxicity within the three following classifications:[15] (1) the toxic substance and its matrix, (2) circumstances of exposure and (3) the subject and its environment (see Figure 1.7). These are considered in the following sections.

Form of the Toxic Substance and its Matrix

Toxicants to which subjects are exposed in the environment or occupationally, particularly through inhalation, may be in several different physical forms. **Gases** are substances such as carbon monoxide in air that are normally in the gaseous state under ambient conditions of temperature and pressure. **Vapors** are gas-phase materials that can evaporate or sublime from liquids or solids. Benzene or naphthalene can exist in the vapor form. **Dusts** are respirable solid particles produced by grinding bulk solids, whereas **fumes** are solid particles from the condensation of vapors, often metals or metal oxides. **Mists** are liquid droplets.

Generally a toxic substance is in solution or mixed with other substances. A substance with which the toxicant is associated (the solvent in which it is dissolved or the solid medium in which it is dispersed) is called the **matrix.** The matrix may have a strong effect upon the toxicity of the toxicant.

Numerous factors may be involved with the toxic substance itself. If the substance is a toxic heavy metal cation, the nature of the anion with which it is associated can be crucial. For example, barium ion, Ba^{2+}, in the form of insoluble barium sulfate, $BaSO_4$, is routinely used as an X-ray-opaque agent in the gastrointestinal tract for diagnostic purposes (barium enema X-ray). This is a safe procedure; however, *soluble* barium salts such as $BaCl_2$ are deadly poisons when introduced into the gastrointestinal tract.

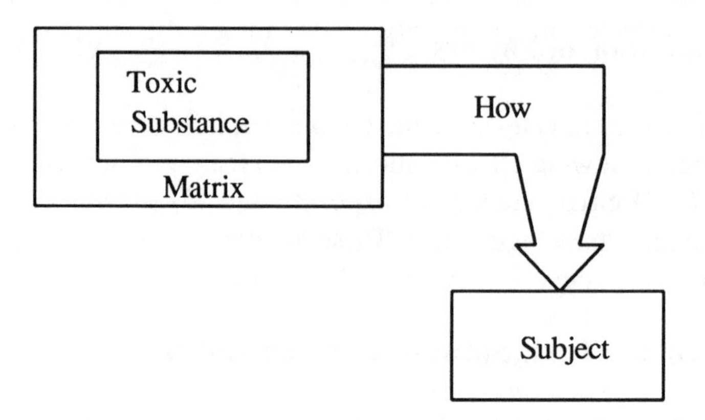

Figure 1.7. Toxicity is influenced by the nature of the toxic substance and its matrix, the subject exposed, and the conditions of exposure.

The pH of the toxic substance can greatly influence its absorption and, therefore, its toxicity. An example of this phenomenon is provided by aspirin, one of the most common causes of poisoning in humans. The chemical name of aspirin is sodium acetylsalicylate, the acidic form of which is acetylsalicylic acid (HAsc), a weak acid that ionizes as follows:

$$\mathrm{HAsc} \rightleftharpoons \mathrm{H^+} + \mathrm{Asc^-} \quad K_a = \frac{[\mathrm{H^+}][\mathrm{Asc^-}]}{[\mathrm{HAsc}]} = 6 \times 10^{-4} \qquad (1.2)$$

The pK_a of HAsc is 3.2 and at a pH substantially below 3.2 most of this acid is in the neutral HAsc form. This neutral form is easily absorbed by the body, especially in the stomach, where the contents have a low pH of about 1. Many other toxic substances exhibit acid-base behavior and pH is a factor in their uptake

Solubility is an obvious factor in determining the toxicity of systemic poisons. These must be soluble in body fluids or converted to a soluble form in the organ or system through which they are introduced into the body. Some insoluble substances that are ingested pass through the gastrointestinal tract without doing harm, whereas they would be quite toxic if they could dissolve in body fluids (see the example of barium sulfate cited above).

As noted at the beginning of this section, the degree of toxicity of a substance may depend on its matrix. The solvent or suspending medium is called the **vehicle**. For laboratory studies of toxicity, several vehicles are commonly used. Among the most common of these are water and aqueous saline solution. Lipid-soluble substances may be dissolved in vegetable oils. Various organic liquids are used as vehicles. Dimethylsulfoxide is a solvent that has some remarkable abilities to carry a solute dissolved in it into the body. The two major classes of vehicles for insoluble substances are the natural gums and synthetic colloidal materials. Examples of the former are tragacanth and acacia, whereas methyl cellulose and carboxymethylcellulose are examples of the latter.

Some drug formulations contain **excipients** that have been added to give a desired consistency or form. In some combinations excipients have a marked influence upon toxicity. **Adjuvants** are excipients that may increase the effect of a toxic substance or enhance the pharmacologic action of a drug. For example, dithiocarbamate fungicides may have their activities increased by the addition of 2-mercaptothiazole.

A variety of materials other than those discussed above may be present in formulations of toxic substances. **Dilutents** increase bulk and mass. Common examples of these are salts, such as calcium carbonate and dicalcium phosphate; carbohydrates, including sucrose and starch; the clay kaolin; and milk solids. Among the **preservatives** used are sodium benzoate, phenylmercuric nitrate and butylated hydroxyanisole (an antioxidant). "Slick" substances such as corn-starch, calcium stearate, and talc act as **lubricants**. Various gums and waxes, starch, gelatin and sucrose are used as **binders**. Gelatin, carnauba wax and shellac are applied as **coating agents**. Cellulose derivatives and starch may be present as **disintegrators** in formulations containing toxicants.

Decomposition may affect the action of a toxic substance. Therefore the stability and storage characteristics of formulations containing toxicants should be considered. A toxic substance may be contaminated with other materials that affect toxicity. Some contaminants may result from decomposition.

Circumstances of Exposure

There are numerous variables related to the ways in which organisms are exposed to toxic substances. One of the most crucial of these, *dose*, was discussed in Section 1.4. Another important factor is the **toxicant concentration**, which may range from the pure substance (100%) down to a very dilute solution of a highly potent poison. Both the **duration** of exposure per exposure incident and the **frequency** of exposure are important. The **rate** of exposure, inversely related to the duration per exposure, and the total time period over which the organism is exposed are both important situational variables. The exposure **site** and **route** also affect toxicity.

It is possible to classify exposures on the basis of acute *vs.* chronic and local *vs.* systemic exposure, giving four general categories.[16] **Acute local** exposure occurs at a specific location over a time period of a few seconds to a few hours and may affect the exposure site, particularly the skin, eyes or mucous membranes. The same parts of the body can be affected by **chronic local** exposure, but the time span may be as long as several years. **Acute systemic** exposure is a brief exposure or exposure to a single dose and occurs with toxicants that can enter the body, such as by inhalation or ingestion, and affect organs such as the liver that are remote from the entry site. **Chronic systemic** exposure differs in that the exposure occurs over a prolonged time period.

The Subject

The first of two major classes of factors in toxicity pertaining to the subject and its environment consists of **factors inherent to the subject**. The most obvious of these is the **taxonomic classification** of the subject, that is, the species and strain. With test animals it is important to consider the **genetic status** of the subjects, including whether they are littermates, half-siblings (different fathers), or the products of inbreeding. Body mass, sex, age, and degree of maturity are all factors in toxicity. **Immunological status** is important. Another area involves the general well-being of the subject. It includes disease and injury, diet, state of hydration, and the subject's "psychological state" as affected by the presence of other species and/or members of the opposite sex, crowding, handling, rest, and activity.

The other of the two major classes of factors related to the subject and its environment consists of **environmental factors**. Among these are ambient atmosphere conditions of temperature, pressure, and humidity, as well as composition of the atmosphere, including the presence of atmospheric pollutants, such as ozone or carbon monoxide. Light and noise and the patterns in which they occur are important. Social and housing (caging) conditions may also influence response of subjects to a toxicant.

LITERATURE CITED

1. Orfila, M. J. B., *Traité des Poisons Tirés des Règnes Minéral, Végétal, et Animal, ou, Toxicologie Générale Considérée sous les Rapports de la Physiologie, de la Pathologie, et de la Médicine Légale*, Crochard, Paris, 1815.

2. Fawcett, Howard, *Hazardous and Toxic Materials*, 2nd ed., John Wiley and Sons, New York, 1988.

3. Hanson, David J., "Industries Straining to File Toxic Release Data by Deadline," *Chemical and Engineering News*, June 20, 1988, pp. 13–16.

4. "Photochemical Smog," Chapter 14 in *Environmental Chemistry*, 4th ed., Stanley E. Manahan, Brooks/Cole Publishing Co., Monterey, CA, 1984.

5. Lipton, Sidney, and Jeremiah Lynch, *Health Hazard Control in The Chemical Process Industry*, John Wiley and Sons, New York, 1987.

6. Williams, Phillip L., and James L. Burson, *Industrial Toxicology*, Van Nostrand Reinhold Co., New York, 1985.

7. Plunkett, E. R., *The Handbook of Industrial Toxicology*, 3rd ed., Chemical Publishing Co., 1987.

8. *National Toxicology Program Fiscal Year 1987 Annual Plan*, U.S. Public Health Service, National Toxicology Program, Research Triangle, NC, 1987.

9. "Toxicological Chemistry in the Environmental Chemistry Curriculum," Stanley E. Manahan, *Preprint Extended*

Abstracts of the Division of Environmental Chemistry, 192nd National Meeting of the American Chemical Society, American Chemical Society, Washington, DC, 1986, p. 223.

10. Walton, Barbara J., and Theodore Mill, "Structure-Activity Relationships in Environmental Toxicology and Chemistry," *Environmental Toxicology and Chemistry*, **7**, 403–404 (1988).

11. Nagamany, Nirmalakhandan, and Richard E. Spreece, "Structure-Activity Relationships," *Environmental Science and Technology*, **22**, 606–615 (1988).

12. Albrecht, W. Klein, Werner Klein, Werner Kördel and Michael Weiss, "Structure-Activity Relationships for Selecting and Setting Priorities for Existing Chemicals — a Computer-Assisted Approach" *Environmental Toxicology and Chemistry*, **7**, 455–467 (1988).

13. "Numbers in Toxicology," Chapter 2 in *Essentials of Toxicology*, 3rd ed., Ted A. Loomis, Lea and Faebiger, Philadelphia, PA, 1978.

14. "Toxicokinetics: The Determinants of Toxicity," Frederick Sperling, Chap. 2 in *Toxicology: Principles and Practice*, Vol. 2, John Wiley and Sons, New York, 1984, pp. 1–18.

15. Doull, John, "Factors Influencing Toxicology," Chapter 5 in *Toxicology: The Basic Science of Poisons*, 2nd ed., John Doull, Curtis D. Klaassen, and Mary O. Amdur, Eds., Macmillan Publishing Co., New York, 1980, pp. 70–83.

16. Haley, Thomas J.,"Toxicology," Section 1 in *Dangerous Properties of Industrial Materials*, 6th ed., N. Irving Sax, Ed., Van Nostrand Reinhold Company, New York, 1984, pp. 1–8.

2

Fundamentals of Chemistry

2.1. CHEMISTRY: THE SCIENCE OF MATTER

Some users of this book will have only a minimal background in chemistry. Although a sound knowledge of chemistry is very helpful in understanding toxicological chemistry, the material in this text can be understood without a detailed background in that discipline. In various parts of the book reference is made to specific aspects of general chemistry, organic chemistry, and biochemistry, which are summarized in this chapter. Readers wanting a somewhat more detailed coverage of these areas are referred to a text on applied chemistry[1] or other standard chemistry texts.

Chemistry is the science of **matter**, which is defined as anything that occupies space and has mass. Matter is the physical material or "stuff" of the universe. Chemistry deals with the composition and properties of matter and the changes that it undergoes. The human body, food, and drinking water are all composed of matter. But matter also occurs as corrosive, flesh-destroying sulfuric acid, deadly poisonous potassium cyanide, and violently explosive nitroglycerin. Therefore, matter is both essential to humankind and potentially very dangerous to it. It is easy to appreciate how a good understanding of the properties of matter is important in dealing effectively and safely with our surroundings.

This chapter is divided into three major parts dealing with chemistry and biochemistry. The first part is a very brief overview of general chemistry, covering at a very fundamental level such concepts as elements, compounds, chemical bonds, chemical formulas, and chemical reactions. Next is discussed organic chemistry, which deals with the unique chemical properties of most of the compounds that contain carbon. Finally, the fundamentals of biochemistry, "the chemistry of life," are discussed. Another crucial area of chemistry, not covered in this text, but required for more sophisticated investigations of toxic and hazardous chemicals and pollutants, is that of analytical chemistry. Many excellent works exist on this rapidly growing subject. A relatively basic coverage of both the classical and instrumental areas of analytical chemistry is contained in the book *Quantitative Chemical Analysis*.[2] Other analytical chemistry books are available that concentrate upon quantitative analysis, instrumental analysis, and more specialized areas, such as atomic spectroscopic analysis of elements, gas chromatography, liquid chromatography, and mass spectrometry.

2.2. ATOMS AND ELEMENTS

As the basic building blocks of matter, **atoms** are extremely important in chemistry. Atoms are almost unimaginably small. An atom consists of two major parts. At the center of each atom is the positively charged **nucleus** containing essentially all of the mass, but occupying virtually none of the volume, of the atom. It is composed of positively charged **protons** and uncharged **neutrons**. The second major portion of the atom — occupying essentially all of the volume, but contributing almost nothing to the mass — is a cloud of negatively charged **electrons**. An uncharged atom has the same number of electrons as protons. The electrons do not move around the nucleus in distinct orbits like planets around the sun. Instead, electrons form clouds of negative charge that may be visualized as spheres, lobes, and rings with indefinite boundaries and exhibiting some of the properties of both particles and waves. Here the atom begins to sound a little like something from science fiction, and a good imagination does help in visualizing atomic structure. Fortunately, for most purposes simplified views of the atom are adequate to explain basic chemical behavior.

Elements and Atomic Number

All atoms having the same number of protons are atoms of the same **element**, the most fundamental kind of substance. For example, all atoms with 6 protons and (and, therefore, 6 electrons) are atoms of the element carbon. The number of protons in the nucleus of each atom of a particular element is the **atomic number** of the element. Atomic numbers are integers ranging from 1 to more than 100. An atomic number denotes a specific element; for example, atomic number 14 specifies silicon.

Each element has a name, such as sodium or nitrogen. In addition to its unique atomic number, each element is denoted by a **chemical symbol** (P for phosphorus, Ca for calcium, or Na for sodium). Some symbols are based upon an element's Latin name, as shown by the example of sodium, for which the Latin name is *natrium*.

Isotopes and Atomic Mass

Atoms of the same element may have different numbers of neutrons in their nuclei. Such atoms are called **isotopes**. The lightest of the elements, hydrogen, atomic number 1, consists of three isotopes. By far the most abundant of the hydrogen isotopes is ordinary hydrogen, which has 1 proton and no neutrons in its nucleus. A much rarer form of hydrogen, deuterium, has 1 proton *and* 1 neutron in the nucleus of each of its atoms. An even more rare isotope is tritium, each atom of which has a nucleus consisting of 1 proton and 2 neutrons. The masses of these isotopes are essentially in the ratio 1:2:3. The nuclei of tritium atoms are unstable and eventually eject a high-energy electron called a beta particle, producing stable isotopes of the element helium. Because of the instability of its nucleus, tritium is a **radioactive isotope**. (There are many radioactive isotopes or **radionuclides** of the various elements, including, for example, the radioactive by-products from nuclear power generation. Radionuclides taken into the body can have serious health effects. Two radioactive elements of concern, radon and radium, are discussed in Chapter 5.)

Most elements have significant fractions of more than one isotope. The average mass of all the atoms of an element is the **atomic mass** (atomic weight) of the element. All atomic masses are expressed relative to carbon-12, the isotope of carbon that contains 6 protons and 6 neutrons in its nucleus. Reference is made to masses of individual atoms in **atomic mass units, μ;** one atomic mass unit is defined as exactly 1/12 the mass of a carbon-12 isotope. Atomic mass units are also called **daltons** in reference to atomic and molecular masses.

2.3. THE PERIODIC TABLE

Central to the organization of all of chemistry is the **periodic table** (Figure 2.1), which lists elements in order of increasing atomic number in a format consistent with the trends and relationships in their chemical properties. Each of the known elements has a box in the periodic table, in which are given the atomic number, chemical symbol, and atomic mass of the element. In the case of chlorine, for example, these are 17, Cl, and 35.453, respectively.

Organization of the Periodic Table

There is significance in the ways in which elements are arranged both vertically and horizontally in the periodic table. Vertical columns of elements are called **groups**. Elements in the A groups (1A and 2A on the left, 3A through 8A on the right) are called **main group** elements, whereas those between main groups 2A and 3A are **transition** elements.The **noble gases**, which are virtually chemically unreactive, are found in the far right column (group 8A) of the periodic table. Elements in the same group tend to have similar chemical properties; this is especially pronounced for groups 1A, 2A, 7A, and 8A. The horizontal rows of elements are called **periods**. The first period has only the two elements hydrogen (H) and helium (He). The second period consists of elements with atomic numbers 3 (lithium) through 10 (neon). The lanthanides and actinides are placed in their own separate periods at the bottom of the table.

Electronic Structure of Atoms

The arrangement of electrons in atoms and the relative energies of the electrons in an atom determine how the atoms behave chemically and, therefore, the chemical properties of each element. It is beyond the scope of this chapter to go into any detail on this subject except to mention several key points. Electrons in atoms occupy **orbitals**, each of which may be occupied by a maximum of 2 electrons. Each orbital is designated by a number (the principal quantum number) and an italicized letter. For orbitals with a particular principle quantum

Active metals		Transition metals										Nonmetals					
1A 1	2A 2	3B 3	4B 4	5B 5	6B 6	7B 7	8	8B 9	10	1B 11	2B 12	3A 13	4A 14	5A 15	6A 16	7A 17	8A 18
1 **H** 1.0079																	2 **He** 4.00260
3 **Li** 6.941	4 **Be** 9.01218											5 **B** 10.81	6 **C** 12.011	7 **N** 14.0067	8 **O** 15.9994	9 **F** 18.998403	10 **Ne** 20.179
11 **Na** 22.98977	12 **Mg** 24.305											13 **Al** 26.98154	14 **Si** 28.0855	15 **P** 30.97376	16 **S** 32.06	17 **Cl** 35.453	18 **Ar** 39.948
19 **K** 39.0983	20 **Ca** 40.078	21 **Sc** 44.9559	22 **Ti** 47.88	23 **V** 50.9415	24 **Cr** 51.996	25 **Mn** 54.9380	26 **Fe** 55.847	27 **Co** 58.9332	28 **Ni** 58.69	29 **Cu** 63.546	30 **Zn** 65.38	31 **Ga** 69.72	32 **Ge** 72.61	33 **As** 74.9216	34 **Se** 78.96	35 **Br** 79.904	36 **Kr** 83.80
37 **Rb** 85.4678	38 **Sr** 87.62	39 **Y** 88.9059	40 **Zr** 91.22	41 **Nb** 92.9064	42 **Mo** 95.94	43 **Tc** (98)	44 **Ru** 101.07	45 **Rh** 102.9055	46 **Pd** 106.42	47 **Ag** 107.8682	48 **Cd** 112.41	49 **In** 114.82	50 **Sn** 118.69	51 **Sb** 121.75	52 **Te** 127.60	53 **I** 126.9045	54 **Xe** 131.29
55 **Cs** 132.9054	56 **Ba** 137.33	57 ***La** 138.9055	72 **Hf** 178.49	73 **Ta** 180.9479	74 **W** 183.85	75 **Re** 186.207	76 **Os** 190.2	77 **Ir** 192.22	78 **Pt** 195.08	79 **Au** 196.9665	80 **Hg** 200.59	81 **Tl** 204.383	82 **Pb** 207.2	83 **Bi** 208.9804	84 **Po** (209)	85 **At** (210)	86 **Rn** (222)
87 **Fr** (223)	88 **Ra** 226.0254	89 **†Ac** 227.0278	104 **Rf** (261)	105 **Ha** (262)	106 **Unh** (263)	107 **Uns** (262)		109 **Une** (266)									

*Lanthanide series	58 **Ce** 140.12	59 **Pr** 140.9077	60 **Nd** 144.24	61 **Pm** (145)	62 **Sm** 150.36	63 **Eu** 151.96	64 **Gd** 157.25	65 **Tb** 158.9254	66 **Dy** 162.50	67 **Ho** 164.9304	68 **Er** 167.26	69 **Tm** 168.9342	70 **Yb** 173.04	71 **Lu** 174.967
†Actinide series	90 **Th** 232.0381	91 **Pa** 231.0359	92 **U** 238.0289	93 **Np** (237)	94 **Pu** (244)	95 **Am** (243)	96 **Cm** (247)	97 **Bk** (247)	98 **Cf** (251)	99 **Es** (252)	100 **Fm** (257)	101 **Md** (258)	102 **No** (259)	103 **Lr** (260)

The larger labels are common American usage. The smaller labels have recently been proposed and are presently under consideration.

Figure 2.1. The periodic table of the elements.

number there is a maximum of 1 *s* orbital, 3 *p* orbitals, 5 *d* orbitals, and 7 *f* orbitals. The **electron configuration** of an atom states the types of orbitals in the atom and the number of electrons in each type. For example, the electron configuration of hydrogen, atomic number 1, is $1s^1$, which simply states that there is 1 electron in the 1*s* orbital of the hydrogen atom. The electron configuration of phosphorus, atomic number 15, is $1s^2 2s^2 2p^6 3s^2 3p^3$, denoting that each phosphorus atom has 2 1*s* electrons, 2 2*s* electrons, 6 2*p* electrons, 2 3*s* electrons, and 3 3*p* electrons. The electron configuration of the noble gas argon, atomic number 18, is $1s^2 2s^2 2p^6 3s^2 3p^6$; note that it is identical to the electron configuration of phosphorus except for the presence of 3 more 3*p* electrons. The writing of electron configurations is simplified by using the chemical symbol of the immediately preceding noble gas in brackets, followed by the appropriate letters and numbers to designate the electrons present beyond the noble gas electron configuration. For example, the electron configuration of element number 23, vanadium, may be written as{Ar}$4s^2 3d^3$, which is equivalent to the much more cumbersome designation of $1s^2 2s^2 2p^6 3s^2 3p^6 4s^2 3d^3$.

For most atoms only the outermost orbitals become involved in the sharing and transfer of electrons involved in chemical bonding and only the outermost electrons take part in this sharing and transfer. In a very general sense, these are the electrons in the electron configuration beyond that of the immediately preceding noble gas in the periodic table. It is convenient for some purposes to show the outermost electrons as dots around the chemical symbol, which is known as a **Lewis symbol**. The Lewis symbols of neon and nitrogen are shown below:

$$:\ddot{\underset{\cdot\cdot}{Ne}}: \qquad :\dot{\underset{\cdot}{N}}\cdot$$

Lewis symbol of neon

Lewis symbol of nitrogen

In the case of neon, 8 electrons (an octet) are shown. This is an especially stable number of outer shell electrons. Atoms of many elements tend to acquire 8 "outer shell" electrons through the sharing, loss, or gain of electrons, which is the basis of the **octet rule** of chemical bonding.

2.4. COMPOUNDS, MOLECULES, AND IONS

Chemical Bonding

Atoms of most elements have a strong tendency to link up with other atoms. Even in the pure elemental form atoms are usually linked together. This can be illustrated with hydrogen. In elemental hydrogen each H atom is joined with another H atom by a **chemical bond** in the form of **molecules** of hydrogen as shown in Figure 2.2. The **chemical formula** of elemental hydrogen is H_2, which shows that each molecule consists of 2 H atoms; such a molecule is called a diatomic molecule. The chemical bond holding the H atoms together in the H_2 molecule is composed of 2 electrons, each contributed by one of the H atoms, and shared between the atoms. The chemical bond can be visualized as a cloud of negative charge in the region between and around the two H nuclei. This cloud is composed of the two rapidly moving electrons which cannot be distinguished by the atom from which they came. A chemical bond in which electrons are shared between atoms is called a **covalent bond**. Another type of bonding where electrons are transferred between atoms is described later in this section.

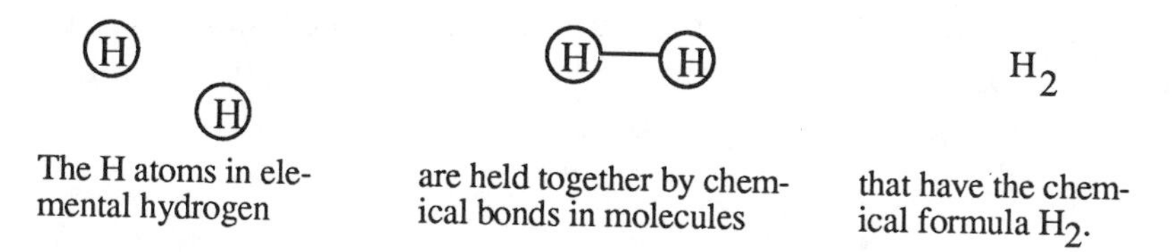

Figure 2.2. Existence of elemental hydrogen as molecules of H_2.

Chemical Compounds

Most substances consist of two or more elements joined by chemical bonds. As an example consider the chemical combination of the elements hydrogen and oxygen shown in Figure 2.3. Oxygen, chemical symbol O, has an atomic number of 8 and an atomic mass of 16.00. Like hydrogen, molecules of oxygen are diatomic and the chemical formula of the most common form of elemental oxygen is O_2. Hydrogen atoms combine with oxygen atoms to form molecules in which 2 H atoms are bonded to 1 O atom to produce a substance with a chemical formula of H_2O (water). A substance such as H_2O that consists of a chemically bonded combination of two or more elements is called a **chemical compound.** In the chemical formula for water the letters H and O are the chemical symbols of the two elements in the compound and the subscript 2 indicates that there are 2 H atoms per O atom. (The absence of a subscript after the O denotes the presence of just 1 O atom in the molecule.) As shown in Figure 2.3, each of the hydrogen atoms in the water molecule is connected to the oxygen atom by a chemical bond. Each of these bonds consists of two electrons shared between the hydrogen and oxygen atoms. For each bond one electron is contributed by the hydrogen and one by oxygen.

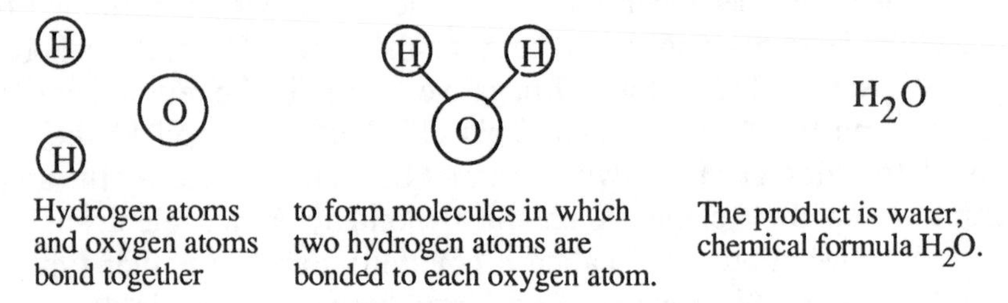

Figure 2.3. Formation of a chemical compound.

In section 2.3 the use of Lewis symbols which show the outer electrons of atoms as dots was mentioned. A similar approach may be used with molecules as shown for water below:

$$\begin{array}{c} \mathrm{H}\cdot\;\cdot\mathrm{H} \\ \cdot\!\!\!\!\!\:\,:\mathrm{O}:\!\!\!\cdot \\ \cdot\,\cdot \end{array}$$ Lewis formula of water

This is the **Lewis formula** of the water molecule. The two dots located between each H and O represent the two electrons in the covalent bond joining these atoms. The four electrons around the oxygen that are not shared with H are non-bonding outer electrons. Note that oxygen has eight (an *octet)* of outer electrons, which was mentioned at the end of Section 2.3 as an especially stable arrangement of electrons for many chemically bound atoms.

Molecular Structure

As implied by Figure 2.3, the atoms and bonds in H_2O form an angle somewhat greater than 90 degrees. The shapes of molecules are referred to as their **molecular geometry**, which is crucial in determining the chemical and toxicological activity of a compound and structure-activity relationships (SAR, Section 1.3).

Ionic Bonds

Molecules of H_2, O_2, and H_2O consist of atoms that are connected by *covalent bonds*, in which electrons are shared between atoms. Another type of chemical bond is illustrated in Figure 2.4. This is the **ionic bond**, in which electrons have been transferred from one atom to another leaving charged species called **ions.** Positively charged ions are called **cations** and negatively charged ions are called **anions.** Ions that make up a solid compound are held together in a **crystalline lattice** consisting of an ordered arrangement of the ions in which each cation is largely surrounded by anions and each anion by cations. The attracting forces of the oppositely charged ions in the crystalline lattice constitute **ionic bonds** in the compound.

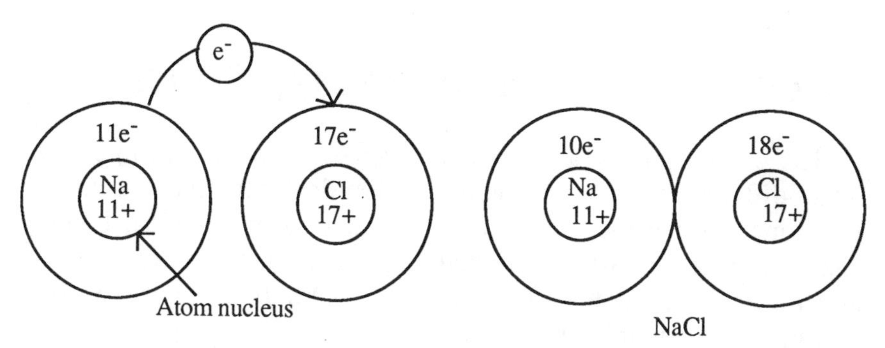

The transfer of an electron from a sodium atom to a chlorine atom

yields Na^+ and Cl^- ions that are bonded together by ionic bonds in the compound NaCl.

Figure 2.4. Ionic bonds are formed by the transfer of electrons and the mutual attraction of oppositely charged ions in a crystalline lattice.

The loss of an electron from the sodium atom as shown above is an example of **oxidation**, and the Na^+ ion product is said to be in the +1 **oxidation state**. In gaining 1 electron in the transfer, the chlorine atom is said to be **reduced.**

Note that in the name of the ionic compound, sodium chloride, the cation, sodium, has the same name as the element from which it was derived whereas the name of the anion, chlor***ide***, has an ending different from that of the element from which it was formed. It is possible for ions to consist of groups of atoms bonded together covalently and having a net charge. Typical of these is the sulfate ion, SO_4^{2-}, which consists of 4 oxygen atoms covalently bonded to a single sulfur (S) atom and having a net electrical charge of -2 for the whole anion.

Molecular Mass

As explained in Section 2.2, the average mass of all atoms of an element is the *atomic mass* of the element. Similarly, molecules of any specific compound have a **molecular mass** (formerly called molecular weight), which is the average mass of all the molecules of the compound. The molecular mass of a compound is calculated by first multiplying the atomic mass of each element by the relative number of atoms of the element and adding all the values obtained for each element in the compound. For example, for H_2O the molecular mass is 2 x 1.0 + 16.0 = 18.0.

2.5. CHEMICAL REACTIONS AND EQUATIONS

The overall process by which substances are converted to chemically different substances is called a **chemical reaction**. Chemical reactions involve the breakage and formation of chemical bonds. Chemical reactions are represented on paper or in computer memory as **chemical equations**. For example, water is produced by the chemical reaction of hydrogen and oxygen:

Hydrogen plus oxygen yields water

Substituting the correct chemical formulas for the names of each reaction participant above leads to

$$H_2 + O_2 \rightarrow H_2O \qquad (2.1)$$

The hydrogen and oxygen to the left of the arrow are called **reactants**.

The water on the right of the arrow is a **product**. The arrow separating the reactants and products is read as "yields." As the equation now stands, it shows a total of 2 H atoms on both sides, but 2 O atoms on the left and 1 on the right, so it is not balanced. It is a **balanced chemical equation** when written as follows:

$$2H_2 + O_2 \rightarrow 2H_2O \qquad (2.2)$$

As with any balanced chemical equation, the chemical formulas in it are correct and it has the same number of each kind of atom on both sides of the arrow, in this case 4 H and 2 O.

2.6. ORGANIC CHEMISTRY

Organic chemistry is that vast branch of chemistry that deals with virtually all carbon-containing compounds, most of which contain carbon atoms covalently bonded to each other. Only a relatively small number of compounds containing carbon are classified as inorganic substances; these include compounds in which carbon is part of simple ions (carbonate, CO_3^{2-}; cyanide, CN^-) and the carbides, such as silicon carbide, SiC. The majority of important industrial compounds, synthetic polymers, biological materials, and most of the toxicants with which this book is concerned are organic compounds. The Chemical Abstracts Service of the American Chemical Society has catalogued about 9 million organic compounds!

It is beyond the scope of this work to teach organic chemistry to a reader unfamiliar with it. However, some basic concepts and definitions can be provided to aid in understanding the organic chemistry material in this book. The reader with no background at all in the subject may want to study a basic organic chemistry textbook. For the non-chemist this is probably best done through one of the several briefer texts used for one-semester organic chemistry courses such as the texts by Hart[3] or McMurry.[4]

Organic Formulas and Structures

In addition to carbon, practically all organic compounds contain hydrogen. The organic compounds composed of just these two elements are called **hydrocarbons**. Many other classes of organic compounds are viewed as derivatives of hydrocarbons in which atoms of other elements have been substituted for one or more hydrogens; for example, replacement of hydrogen by chlorine yields organochlorine compounds, which are of great industrial and toxicological importance.

Molecular formulas of organic compounds give the number of each kind of atom in a molecule of a compound. For example, the molecular formula C_8H_{18} denotes a hydrocarbon that may be called octane, where the "oct–" prefix indicates that the compound has eight carbon atoms in each molecule. However, because of the ability of carbon atoms to bond to each other in chains and branched chains, there are several different hydrocarbons with the C_8H_{18} molecular formula as shown in Figure 2.5.

```
                                                  H
                                                  |
                                                H-C-H
  H H H H H H H H                               H |  H H H H
  | | | | | | | |                               | |  | | | |
H-C-C-C-C-C-C-C-C-H                           H-C-C-C-C-C-C-H
  | | | | | | | |                               | | | | | |
  H H H H H H H H                               H H H H | H
                                                        |
     n-Octane             2,5-Dimethylhexane          H-C-H
                                                        |
                                                        H

        H
        |
      H-C-H
      H |  H H H
      | |  | | |
    H-C-C-C-C-C-H
      | | | | |
      H H | H H
        H-C-H
          |             2-Methyl-,3-ethylpentane
        H-C-H
          |
          H
```

Figure 2.5. Structural formulas of three isomers of octane, an alkane hydrocarbon with the molecular formula C_8H_{18}.

The formulas in Figure 2.5 are **structural formulas** that show the relative placement of atoms and groups of atoms, thereby providing much more information than do molecular formulas. Each of the three compounds illustrated in the figure has somewhat different chemical, physical, and toxicological properties, despite their identical molecular formulas. Compounds such as these having the same molecular but different structural formulas are called **structural isomers.**

Structural formulas can be abbreviated in several ways. For example, the structural formula of 2,5-dimethylhexane can be written as $CH_3CH(CH_3)CH_2CH_2CH(CH_3)CH_3$, where the CH_3 (methyl) groups are placed in parentheses to show that they are side-chain substituent groups attached to the immediately preceding carbon atom in the longest (6-carbon hexane) chain.

In some cases different isomers occur in which the atoms in an organic molecule occur in the same order, but occupy different locations in space. The easiest such case to understand may be illustrated with an example of an **alkene**, a type of hydrocarbon containing a **double bond** consisting of *four* electrons shared between two carbon atoms, a bond which is illustrated by two parallel lines, =. Two alkenes with the same molecular formula, C_4H_8, are shown in Figure 2.6. To understand why these are two different compounds, it must be realized that the two atoms joined by the double bond cannot rotate relative to each other without breaking the bond. In the case of the *cis* isomer, the two end methyl groups are "on the same side," whereas with the *trans* isomer they are opposite. This kind of isomerism is called ***cis-trans*** isomerism. It is a special case of **stereoisomerism** that involves the spacial arrangements of atoms that occur in the same order in a molecule.

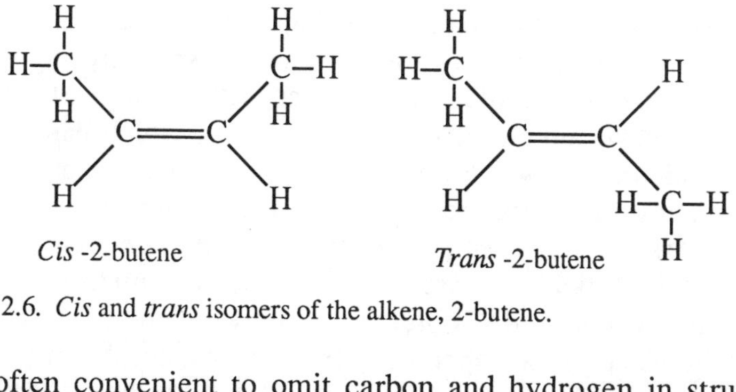

Cis -2-butene

Trans -2-butene

Figure 2.6. *Cis* and *trans* isomers of the alkene, 2-butene.

It is often convenient to omit carbon and hydrogen in structural formulas in which the skeleton of the organic molecule is shown by lines and figures such as hexagons. In such a structure each "corner" denotes the location of a carbon atom and a carbon atom is understood to be present at the end of each line. Furthermore, a carbon atom at a corner to which two lines are drawn (each representing two electrons in a covalent bond) is understood to have *two* H atoms attached to it, one H atom is attached to a carbon represented by the intersection of three lines, and none where four lines intersect. These relationships are illustrated in Figure 2.7, which illustrates some compounds for which structural formulas were given previously in this section.

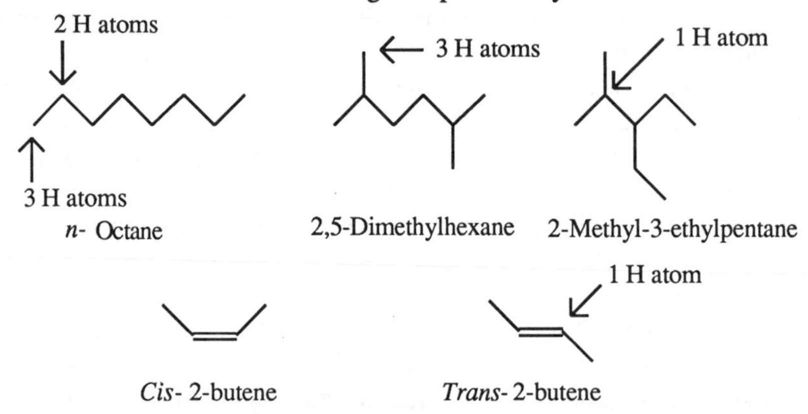

Figure 2.7. Representation of structural formulas with lines (see more detailed formulas in Figures 2.5 and 2.6). A carbon atom is understood to be at each corner. The numbers of hydrogen atoms attached to carbons at several specific locations are shown by arrows.

Molecular Geometry

As pointed out in Section 2.4, the shapes of molecules are covered under the topic of molecular geometry and are especially important in areas such as how molecules interact with biological systems. Molecules have three-dimensional structures, which can be challenging to represent on paper. However, much can be done to indicate three-dimensional structures by using lines in which (1) those of normal thickness represent parts of the structure (bonds) in the plane of the paper, (2) bold lines are viewed as extending up from the paper toward the reader, and (3) broken lines are regarded as extending below the plane of the paper away from the reader. As an example, briefly consider dichloromethane (methylene chloride), CH_2Cl_2, an important organochloride solvent and extractant. The four atoms bonded to the central C atom do not lie in a plane, but can be visualized as comprising the corners of a tetrahedron in the form of a four-sided pyramid in which each face is a triangle. The structural formula on the left in Figure 2.8 implies none of this. In the structural formula on the right, however, the broken lines indicate that the two hydrogen atoms lie "back" and the two Cl atoms "forward." This simple molecule could be rotated, of course, so that different perspectives are presented to the reader.

Figure 2.8. Structural formulas of dichloromethane, CH_2Cl_2. The formula on the right provides a three-dimensional representation in which the two Cl atoms are represented as standing out from the plane of the paper and the two H atoms are projected rearward.

Aromatic Organic Compounds

Figure 2.9 represents the compound benzene, C_6H_6. It is the most prominent member of a class of organic compounds called **aromatic compounds** or **arenes** that are characterized by ring structures and delocalized clouds of so-called π (pi, pronounced "pie") electrons. These compounds are very important in all areas of organic chemistry and in toxicological chemistry.

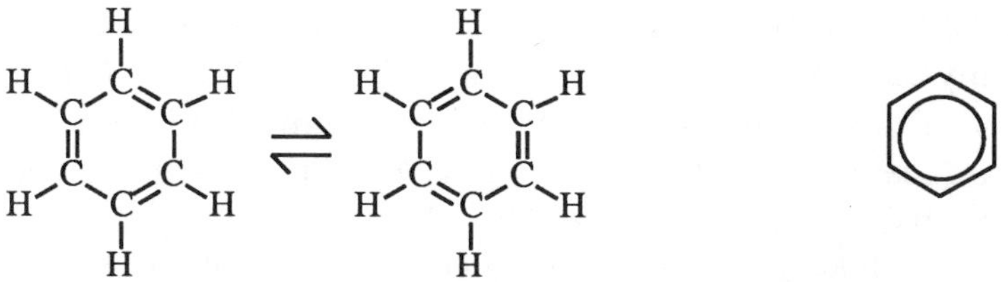

Figure 2.9. The aromatic benzene molecule represented by two equivalent structures each of which has 3 equivalent bonds (left). This representation is not consistent with the properties of the compound. It is shown more properly as a hexagon with a circle in it (right). In this structure a carbon atom is understood to be at each corner with one H atom attached, unless some other atom is shown specifically attached to the C atom.

In an oversimplified sense, the structure of benzene can be visualized as resonating between the two equivalent structures shown on the left in Figure 2.9 by the shifting of electrons in chemical bonds. This structure can be shown more simply by a hexagon with a circle in it (see caption to Figure 2.9).

Many aromatic compounds, including the toxicologically significant class of polycyclic aromatic hydrocarbons (PAH), consist of benzene rings condensed together. The simplest example is naphthalene (below), molecular formula $C_{10}H_8$. In this structure each of the 8 "outside" corners represents a C atom with 1 H attached and each of the 2 "inside" corners stands for a C atom with no H atoms attached.

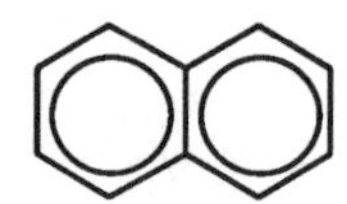

Naphthalene

Naming Organic Compounds

Many important commercial organic compounds have common names and all documented organic compounds have **systematic names** that describe the compound structures. The naming of organic compounds is covered under the topic of **organic nomenclature**. It is not possible to cover organic nomenclature in any detail here, but a few examples should be helpful in understanding the nomenclature of organic compounds in this book.

The names of organic chemicals are based upon the names of groups attached to numbered positions on carbon chains and rings. In Figure 2.5 a "straight-chain," 8-carbon hydrocarbon was shown. (Actually the chains are kinked because of the angles that C-C bonds form with carbon atoms as illustrated in Figure 2.7.) The hydrocarbon in question is called *n*-octane where "*n*" denotes that it is a straight-chain molecule without side chains and octane is the name assigned to an 8-carbon hydrocarbon without any multiple bonds. The second compound shown in Figure 2.5 has 6 carbon atoms in its longest continuous chain of carbon atoms, so its root name is hexane, where "hex" denotes 6. Two methyl groups are attached to this chain at carbons 2 and 5, so the compound is called 2,5-dimethylhexane, where "di" stands for two. The third compound shown in the figure has 5 carbon atoms in its longest continuous carbon chain with a methyl group substituted on the second carbon of the chain and an ethyl group (C_2H_5) on the third carbon atom, so the name is 2-methyl-3-ethylpentane.

The nomenclature of a ring compound can be illustrated by the two derivatives of benzene shown in Figure 2.10. The base compound

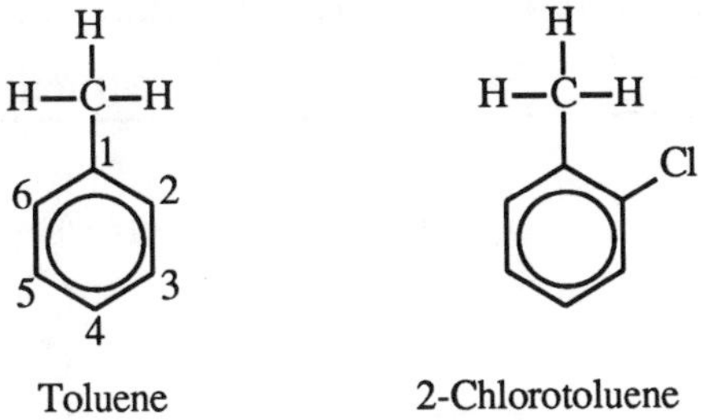

Toluene 2-Chlorotoluene

Figure 2.10. Structural formulas of toluene, C_7H_8 and 2-chlorotoluene, C_7H_7Cl.

is toluene, which has a benzene ring in which one of the H atoms has been replaced by a methyl group. The 6 C atoms on the ring are numbered consecutively, starting with the one on which the methyl group is attached. In the compound on the right, the H atom attached to the second carbon atom in the ring has been replaced with a chlorine atom, so the compound is named 2-chlorotoluene. The first compound could also be named methylbenzene and the second 1-methyl-2-chlorobenzene. In older nomenclature still frequently encountered, positions closest to the position of the major substituent group (2 and 6) are called *ortho*, the next two positions (3 and 5) are called *meta*, and the farthest position (4) is called *para*. Using this system, the chloro derivative in Figure 2.10 would be called *ortho*-chlorotoluene.

Functional Groups

In addition to carbon atoms linked together in straight chains, branched chains, and rings, usually with H atoms attached, organic molecules may contain groups of atoms or bonds known as **functional groups** (Table 2.1). These form a major basis for classifying organic compounds. Although most functional groups contain atoms other than carbon or hydrogen, the C=C double bond (see Figure 2.6) is considered a functional group that defines the class of hydrocarbons called *alkenes*. The carbon-carbon triple bond, such as the multiple bond in acetylene,

$$H-C{\equiv}C-H \qquad \text{Acetylene}$$

Table 2.1 Some Common Functional Groups

Class of compound	Example	Structural formula[1]
Alcohol	Methanol	$H-C(H)(H)-OH$
Ether	Dimethyl ether	$H-C(H)(H)-O-C(H)(H)-H$
Aldehyde	Propionaldehyde	$H-C(H)(H)-C(H)(H)-C(=O)-H$
Amine	Methylamine	$H-C(H)(H)-N(H)(H)$
Nitro compounds	Nitromethane	$H-C(H)(H)-NO_2$
Heterocyclic nitrogen compounds	Pyridine	N
Nitroso compounds	α–Nitroso–β–naphthol	NO, OH
Sulfonic acids	Benzenesulfonic acid	$-S(=O)(=O)-OH$
Organohalides	1,1–Dichloro-ethane	$Cl-C(H)(H)-C(H)(H)-Cl$

[1] Functional groups are outlined with dashed lines.

(consisting of 6 shared electrons) is present in **alkyne** hydrocarbons. In addition to the functional groups shown in Table 2.1, other examples are noted throughout the text in the chapters dealing with classes of organic chemicals.

Organic Polymers

Much of the organic chemistry industry is based upon the manufacture of **polymers** consisting of high-molecular-mass molecules formed by the joining together of individual molecules called **monomers**. The reaction for the formation of a typical polymer, polystyrene, is the following:

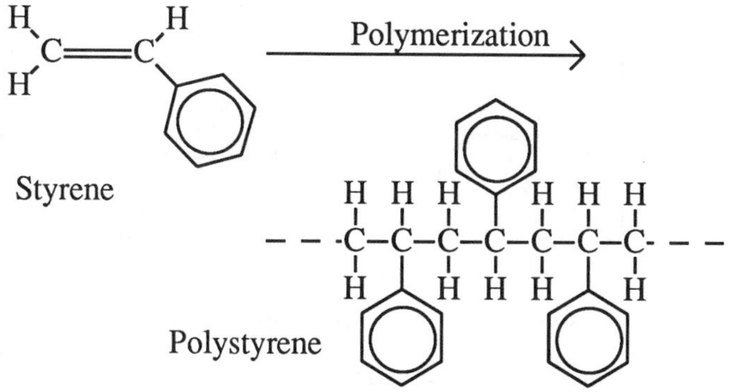

Styrene —Polymerization→ Polystyrene (2.3)

Plastics, textiles, and many other products are polymers. Many of the monomers used to manufacture polymers are toxic, reactive and flammable. Furthermore, many plastics contain environmentally persistent **plasticizers** required to give them desired properties (see phthalates in Chapter 9). When burned, some polymers evolve noxious gases, such as HCl gas produced in the incineration of polyvinylchloride. Because of factors such as these and the massive amounts of polymers produced, their toxicological chemistry, and that of the monomers used to make them, is of particular concern.

2.7. BIOCHEMISTRY AND THE CELL

Biochemistry is the science that deals with chemical processes and materials in living systems. As such, it is basic to toxicology.[5] Biochemistry is introduced very briefly here. Aspects of the discipline are covered in later chapters, particularly Chapter 4, "Biochemical Action and Transformation of Toxicants."

The Cell

The focal point of biochemistry and biochemical aspects of toxicants is the **cell**, the basic building block of living systems where most life processes are carried out. The type of cell in higher forms of animal and plant life (multicelled organisms) is the **eukaryotic cell**[6] shown in Figure 2.11.

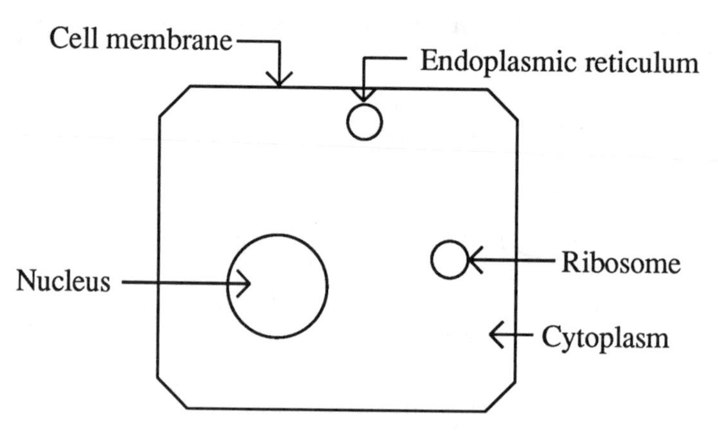

Figure 2.11. Some features of the eukaryotic cell.

Eukaryotic cells are quite complex with numerous kinds of bodies, organelles, and membranes. Each is enclosed by a **cell membrane** and contains a **nucleus**, both of which are discussed below. Much of the material filling the cell is **cytoplasm**. Cell **mitochondria** are involved with energy conversion and utilization. **Ribosomes** are key bodies in protein synthesis. The **endoplasmic reticulum** is very important in the metabolism of some toxicants.

Cell Membrane and Cell Nucleus

Two parts of the cell are so important to toxicology that they are discussed briefly here. The first of these is the *cell membrane*, which encloses the cell and regulates the passage of materials into and out of the cell. These materials include nutrients, metabolic products (including wastes), ions, and lipid-soluble ("fat-soluble") substances.

The passage of toxicants and their products through the cell membrane is especially important in toxicology. The cell membrane is actually a rather complicated structure consisting largely of a stacked arrangement of molecules called phospholipids, which have their hydrophilic ("water-loving") heads on the outside and inside surfaces of the membrane and their hydrophobic ("water-repelling") tails in the inside of the membrane. Included also in the cell membrane are bodies of proteins (see below), which are involved in the passage of substances through the membrane. Cell membrane damage is a major mode of toxic action.

The *nucleus* of the cell contains **chromosomes** that regulate cell division and the synthesis of proteins.[7] The chromosomes contain **genes** composed of **deoxyribonucleic acid, DNA**. This is the basic material of cell reproduction and heredity, widely publicized in recent years in connection with emerging technologies involving recombinant DNA and genetic engineering. Because of its crucial roles in heredity, cell reproduction, and protein synthesis, DNA damage and alteration by toxic substances can lead to mutations, cancer, birth defects, defective function of immune systems, and other manifestations of poisoning.

Proteins, Carbohydrates, and Lipids

Proteins, carbohydrates, and lipids are the three major types of life substances that are synthesized in large quantities in cells and are also broken down metabolically to provide energy and materials for additional synthesis. Many toxicants interfere with these vital processes. Furthermore, some toxic substances alter proteins (such as by binding to proteinaceous enzymes), carbohydrates, and lipids with detrimental effects to organisms.

Proteins

Proteins are polymers (see previous discussion) of amino acids, such as glycine and phenylalanine (Figure 2.12), that make up most of the cytoplasm inside the cell, consitute enzymes, and have other essential functions in living systems. The amino acids in proteins are joined at **peptide linkages** (Figure 2.12). The order of amino acids in a protein molecule determines its primary structure. In very general terms, the shapes of huge protein molecules define their more advanced structural features, which are called secondary, tertiary, and quaternary structures. Protein structure is very important in areas such as the ability of the body's immune system to recognize specific proteins and in the action of proteinaceous enzymes.

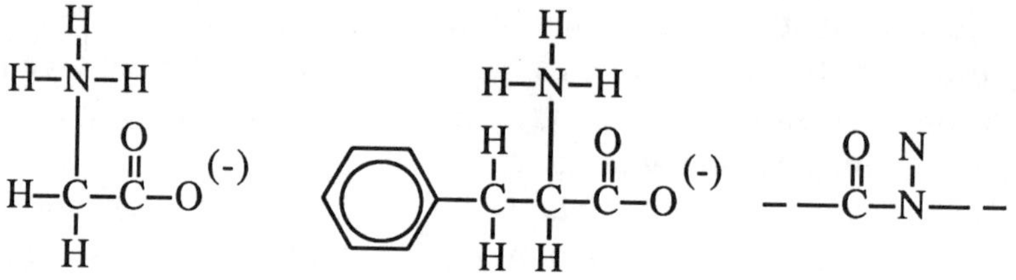

Glycine Phenylalanine Peptide linkage

Figure 2.12. Two typical amino acids and the peptide linkage in proteins.

Carbohydrates

Carbohydrates have the approximate simple formula CH_2O and include a diverse range of substances. Some, like glucose,

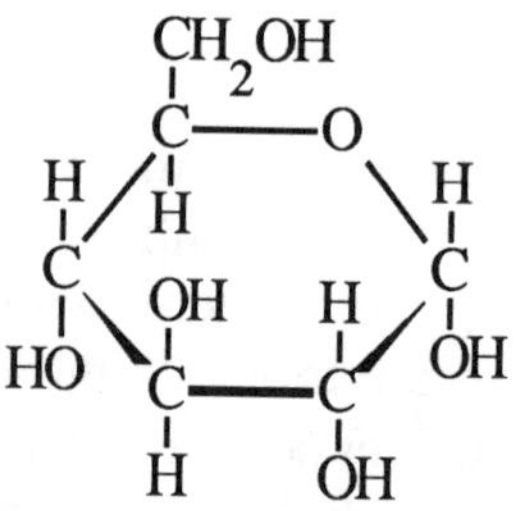

Glucose molecule

are simple sugars, whereas others are polymers of high molecular mass such as starch and cellulose. Carbohydrates are used in living systems primarily for storing and transferring energy. A simple sugar such as glucose is called a **monosaccharide. Disaccharides** are composed of molecules consisting of two simple sugar molecules linked together; a molecule of H_2O is lost for each linkage formed. **Polysaccharides** are polymers consisting of numerous simple sugar molecules.

Lipids

Lipids constitute a diverse class of biological materials that can be extracted from plant or animal matter by organic solvents, such as chloroform, ether, or benzene. The most common lipids are fats and oils, which have the general structure shown in Figure 2.13. Waxes, cholesterol, and some vitamins and hormones are lipids. Lipids are especially important in toxicology because of their ability to dissolve and store toxicants that are poorly soluble in water. Most toxic organic compounds are lipid-soluble.

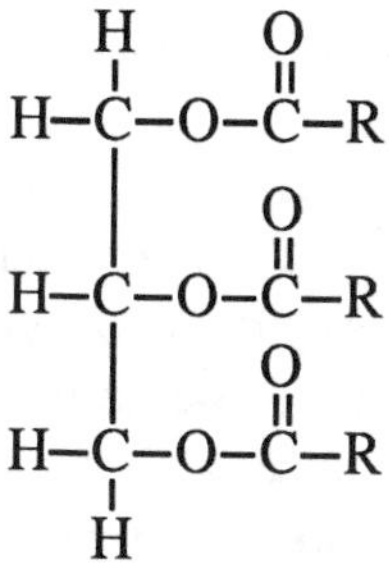

```
   H     O
   |     ||
H—C—O—C—R
   |     O
   |     ||
H—C—O—C—R
   |     O
   |     ||
H—C—O—C—R
   |
   H
```

Figure 2.13. General formula of fats and oils. The R group is a hydrocarbon chain, such as $-(CH_2)_{16}CH_3$.

Enzymes

Enzymes are biological catalysts that enable biochemical processes to occur. Most enzymes consist of protein molecules with very specific structures. The defined shape of an enzyme gives it **specificity** to interact with a particular substance or class of substances called a **substrate**, as illustrated in Figure 2.14. The product of the reaction of an enzyme and substrate is an enzyme-substrate complex that may produce products different from the substrate when it comes apart, regenerating unchanged enzyme that can catalyze additional reactions.

Many of the functions that enzymes perform can be placed in several different categories. For example, **hydrolyzing enzymes** catalyze the addition of H_2O to polymeric biological compounds such as carbohydrates, causing these compounds to break down to lower molecular mass products. Enzymes are named for what they do. For example, the stomach (gastric) enzyme that breaks down proteins is called *gastric proteinase.*

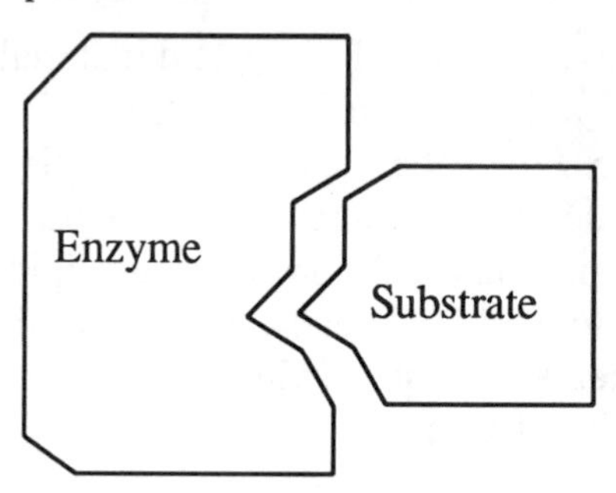

Figure 2.14. Structure of an enzyme that enables it to fit a specific substrate.

The mode of toxic action of many toxicants is through the alteration or destruction of enzymes with the result that the enzymes no longer function or they work in detrimental ways. Among the toxicants that damage enzymes are cyanide, heavy metals and formaldehyde. Some of these substances cause the enzyme structure to unravel, thus changing its shape. Heavy metals attached to the enzyme surface alter its shape so that it no longer works properly.

LITERATURE CITED

1. Manahan, Stanley E., *General Applied Chemistry*, 2nd ed., Brooks/Cole Publishing Co., Pacific Grove, CA, 1982.

2. Manahan, Stanley E., *Quantitative Chemical Analysis*, Brooks/Cole Publishing Co., Monterey, CA, 1986.

3. Hart, Harold, *Organic Chemistry — A Short Course*, 7th ed., Houghton Mifflin Company, Boston, 1987.

4. McMurry, John, *Fundamentals of Organic Chemistry*, Brooks/Cole Publishing Co., Pacific Grove, CA, 1986.

5. Hodgson, Ernest and Frank E. Guthrie, Eds., *Introduction to Biochemical Toxicology*, Elsevier, New York, 1980.

6. Sears, Curtis C., and Conrad L. Stanitski, *Chemistry for*

Health-Related Sciences, 2nd ed., Prentice-Hall, Inc., Englewood Cliffs, NJ, 1983.

7. Dickson, T. R., *Introduction to Chemistry*, 4th ed., John Wiley and Sons, New York, 1983.

3

Exposure and Effects of Toxic Substances

3.1. INTRODUCTION

This chapter deals with the routes of exposure and clinically observable effects of toxic substances. The information is presented primarily from the viewpoint of human exposure and readily observed detrimental effects of toxic substances on humans. To a somewhat lesser extent this material applies to other mammals, especially those used as test organisms. It should be kept in mind that many of the same general principles discussed apply also to other of the more complex organisms, such as fish and even plants.

Although LD_{50} (See Section 1.4) is often the first parameter to come to mind in discussing degrees of toxicity, mortality is usually not a good parameter for toxicity measurement. Much more widespread than fatal poisoning, and certainly more subtle, are various manifestations of morbidity ("unhealthiness"). As discussed in this chapter, there are many ways in which morbidity is manifested. Some of these, such as effects on vital signs, are obvious. Others, such as some kinds of immune system impairment, require sophisticated tests to be observed. Various factors must be considered, such as minimum dose or the latency period (often measured in years for humans) for an observable response to be observed. Furthermore, it is important to distinguish **acute toxicity**, which has an effect soon after exposure, and **chronic toxicity**, which has a long latency period.

3.2 ROUTES OF EXPOSURE

The major routes of accidental or intentional exposure to toxicants by humans and other animals are the skin (percutaneous route), the lungs (inhalation, respiration, pulmonary route), and the mouth (oral route); minor routes of exposure are rectal, vaginal, and parenteral (intravenous or intramuscular, a common means for the administration of drugs or toxic substances in test subjects).[1] The way that a toxic substance is introduced into the complex system of an organism is strongly dependent upon the physical and chemical properties of the substance.[2] The pulmonary system is most likely to take in toxic gases or very fine, respirable solid or liquid particles. In other than a respirable form, a solid usually enters the body orally. Absorption through the skin is most likely for liquids, solutes in solution, and semisolids, such as sludges.

The defensive barriers that a toxicant may encounter vary with the route of exposure. For example, elemental mercury is readily absorbed, often with devastating effect, through the alveoli in the lungs much more readily than through the skin or gastrointestinal tract. Most test exposures to animals are through ingestion or gavage. Pulmonary exposure is often favored with subjects that may exhibit refractory behavior when noxious chemicals are administered by means requiring a degree of cooperation from the subject. Intravenous injection may be chosen for deliberate exposure when it is necessary to know the concentration and effect of a xenobiotic substance in the blood. However, pathways used experimentally that are almost certain not to be significant in accidental exposures can give misleading results when they avoid the body's natural defense mechanisms.

In discussing exposure sites for toxicants it is useful to consider the major routes and sites of exposure, distribution, and elimination of toxicants in the body as shown in Chapter 1, Figure 1.5.

3.3. PERCUTANEOUS EXPOSURE

Toxicants can enter the skin through epidermal cells, sebaceous gland cells, or hair follicles.[3] By far the greatest area of the skin is composed of the epidermal cell layer, and most toxicants absorbed through the skin do so through epidermal cells. Despite their much smaller total areas, however, the cells in the follicular walls and in sebaceous glands are much more permeable than epidermal cells.

Skin Permeability

Figure 3.1 illustrates the absorption of a toxic substance through the skin and its entry into the circulatory system, where it may be distributed through the body. Often the skin suffers little or no harm at the site of entry of systemic poisons, which may act with devastating effects upon receptors far from the location of absorption.

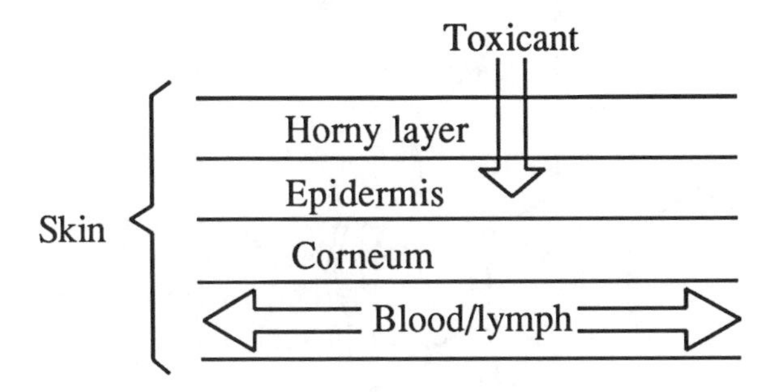

Figure 3.1. Absorption of a toxic substance through the skin.

The permeability of the skin to a toxic substance is a function of both the substance and the skin. The permeability of the skin varies with both the location and the species that penetrates it. Tests of two organophosphorus compounds have shown variations of up to 8:1 for the percutaneous absorption of the *two different compounds* by the same species and as much as 33:1 for the absorption of the *same compound* by two different species of mammals.[4] In order to penetrate the skin significantly, a substance must be a liquid or gas or significantly soluble in water or organic solvents. In general, nonpolar, lipid-soluble substances traverse skin more readily than do ionic species. Substances that penetrate skin easily include lipid-soluble endogenous substances (hormones, vitamins D and K) as well as a number of xenobiotic compounds. Examples of these are shown in Figure 3.2. Some military poisons, such as the "nerve gas" Sarin (see Chapter 13), permeate the skin very readily, which greatly adds to their hazards. In addition to the rate of transport through the skin, an additional factor that influences toxicity via the percutaneous route is the blood flow at the site of exposure.

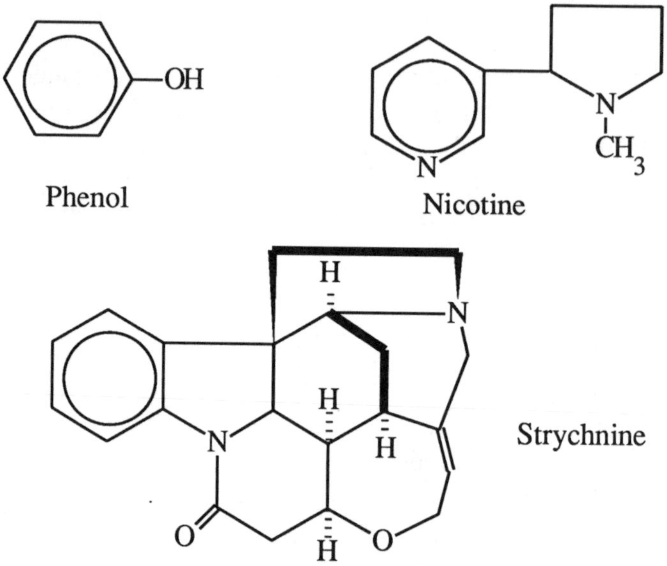

Phenol

Nicotine

Strychnine

Figure 3.2. Examples of xenobiotic compounds that may be absorbed through the skin.

Barriers to Skin Absorption

The major barrier to dermal absorption of toxicants is the **stratum corneum** or horny layer (see Figure 3.1). The permeability of skin is inversely proportional to the thickness of this layer, which varies by location on the body in the order soles and palms > abdomen, back, legs, arms > genital (perineal) area. Evidence of the susceptibility of the genital area to absorption of toxic substances is to be found in accounts of the high incidence of cancer of the scrotum among chimney sweeps in London described by Sir Percival Pott, Surgeon General of Britain during the reign of King George III. The cancer-causing agent was coal tar condensed in chimneys. This material was more readily absorbed through the skin in the genital areas than elsewhere leading to a high incidence of scrotal cancer. (The chimney sweeps had little appreciation of basic hygienic practices, such as bathing and regular changes of underclothing.) Breaks in epidermis due to laceration, abrasion, or irritation increase the permeability, as do inflammation and higher degrees of skin hydration.

Measurement of Dermal Toxicant Uptake

There are two principal methods for determining the susceptibility of skin to penetration by toxicants. The first of these is measurement of the dose of the substance received by the organism using chemical analysis, radiochemical analysis of radioisotope-labelled substances, or observation of clinical symptoms. Secondly, the amount of substance remaining at the site of administration may be measured. The latter approach requires control of nonabsorptive losses of the substance, such as those that occur by evaporation.

3.4. PULMONARY EXPOSURE

The pulmonary system is the site of entry for numerous toxicants. Examples of toxic substances inhaled by human lungs include fly ash and ozone from polluted atmospheres, vapors of volatile chemicals used in the workplace, tobacco smoke, radioactive radon gas, and vapors from paints, varnishes, and synthetic materials used for building construction.

The major function of the lungs is to exchange gases between the bloodstream and the air in the lungs. This includes especially the absorption of oxygen by the blood and the loss of carbon dioxide. Gas exchange occurs in a vast number of alveoli in the lungs where a tissue the thickness of only one cell separates blood from air. The thin, fragile nature of this tissue makes the lungs especially susceptible to absorption of toxicants and to direct damage from toxic substances. Furthermore, the respiratory route enables toxicants entering the body to bypass organs that have a screening effect (the liver is the major "screening organ" in the body and it acts to detoxify numerous toxic substances). These toxicants can enter the bloodstream directly and be transported quickly to receptor sites with minimum intervention by the body's defense mechanisms.

As illustrated in Figure 3.3, there are several parts of the pulmonary system that can be affected by toxic substances. The upper respiratory tract consisting of the nose, throat, trachea, and bronchi retains larger particles that are inhaled. The retained particles may cause upper respiratory tract irritation. Cilia, which are small hair-like appendages in the upper respiratory tract, move with a sweeping motion to remove captured particles. These substances are transported to the throat from which they may enter the gastrointestinal tract and be absorbed by the body. Gases such as ammonia (NH_3) and hydrogen chloride (HCl) that are very soluble in water are also removed from air predominantly in the upper respiratory tract and may be very irritating to tissue in that region.

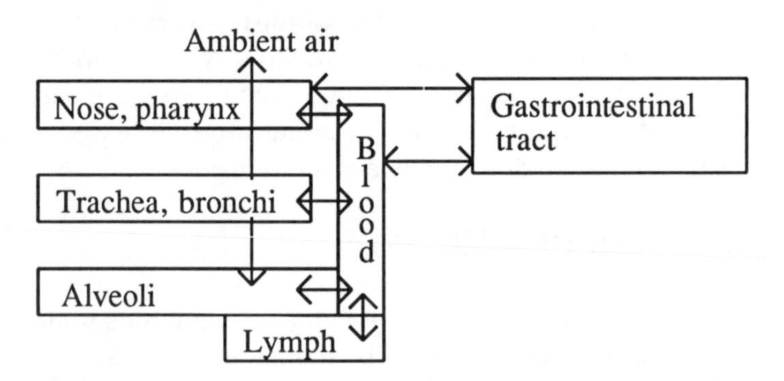

Figure 3.3. Pathways of toxicants in the respiratory system.

3.5. GASTROINTESTINAL TRACT

The gastrointestinal tract may be regarded as a tube through the body from the mouth to the anus, the contents of which are external to the rest of the organism system. Therefore, any systemic effect of a toxicant requires its absorption through the mucosal cells that line the inside of the gastrointestinal tract. Caustic chemicals can destroy or damage the internal surface of the tract and are viewed as nonkinetic poisons (see Section 1.9).

Mouth, Esophagus, and Stomach

Most substances are not readily absorbed in the mouth or esophagus; one of several exceptions is nitroglycerin administered for certain heart disfunctions and absorbed if left in contact with oral tissue. The stomach is the first part of the gastrointestinal tract where substantial absorption and translocation to other parts of the body may take place. The stomach is unique because of its high content of HCl and consequent low pH (about 1.0). Therefore, some substances that are ionic at pH near 7 and above are neutral in the stomach so that they readily traverse the stomach walls (see the example of acetosalicylic acid in Section 1.11). In some cases absorption is affected by stomach contents other than HCl. These include food particles, gastric mucin, gastric lipase, and pepsin.

Intestines

The small intestine is effective in the absorption and translocation of toxicants. The pH of the contents of the small intestine is close to neutral, so that weak bases that are charged (HB^+) in the acidic environment of the stomach are uncharged (B) and absorbable in the intestine. The small intestine has a large surface area favoring absorption. Intestinal contents are moved through the intestinal tract by peristalsis. This has a mixing action on the contents and enables absorption to occur the length of the intestine. Some toxicants slow down or stop peristalsis (paralytic ileus), thereby slowing the absorption of the toxicant itself.

The Intestinal Tract and the Liver

The intestine/blood/liver/bile loop constitutes the **enterohepatic circulation** system (see Figure 3.4). A substance absorbed through the intestines goes either directly to the lymphatic system or to the **portal**

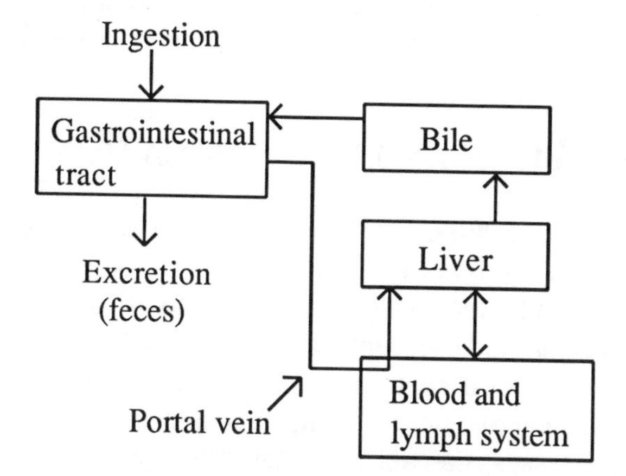

Figure 3.4. Representation of enterohepatic circulation.

circulatory system. The latter carries blood to the portal vein that goes directly to the liver. The liver serves as a screening organ for xenobiotics (see Section 3.4) and secretes these substances or a metabolic product of them back to the intestines. For some substances there are mechanisms of active excretion into the bile in which the substances are concentrated by 1–3 orders of magnitude over levels in the blood. Other substances enter the bile from blood simply by diffusion.

3.6. PHASES OF TOXICITY

Having examined the routes by which toxicants enter the body, it is now appropriate to consider what happens to them in the body and what their effects are. The action of a toxic substance can be divided into two major phases[5] as illustrated in Figure 3.5. The **kinetic phase** involves absorption, metabolism, temporary storage, distribution, and to a certain extent excretion of the toxicant or its precursor compound called the **protoxicant**. In the most favorable scenario for an organism, a toxicant is absorbed, detoxified by metabolic processes, and excreted with no harm resulting. In the least favorable case, a protoxicant that is not itself toxic is absorbed and converted to a toxic metabolic product which is transported to a location where it has a detrimental effect. The **dynamic phase** is divided as follows: (1) the toxicant reacts with a receptor or target organ[6] in the **primary reaction** step, (2) there is a biochemical response and (3) physiological and/or behavioral manifestations of the effect of the toxicant occur.

3.7. TOXIFICATION AND DETOXIFICATION

As shown for the kinetic phase in Figure 3.5, a xenobiotic substance may be (1) detoxified by metabolic processes and eliminated from the body, (2) made more toxic (toxified) by metabolic processes and distributed to receptors, or (3) passed on to receptors as a metabolically unmodified toxicant. A metabolically unmodified toxicant is called an **active parent compound** and a substance modified by metabolic processes is an **active metabolite**. Both types of species may be involved in the dynamic phase.[7]

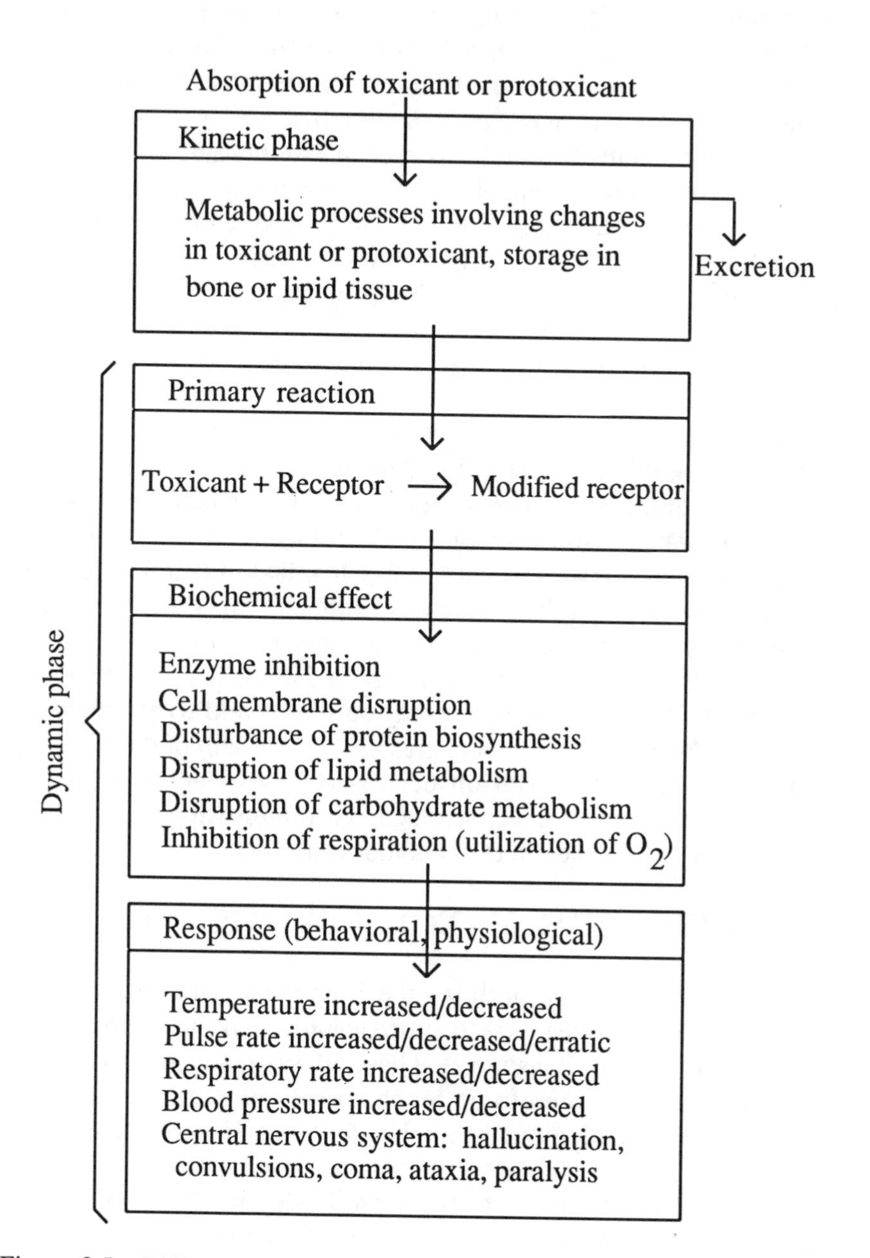

Figure 3.5. Different major steps in the overall process leading to a toxic response.

During the kinetic phase an active parent compound can be present in blood, liver, or **extrahepatic tissue** (nonliver tissue), and in the latter two it may be converted to an inactive metabolite or metabolites. An inactive parent metabolite may produce a toxic metabolite or metabolites in the liver or in extrahepatic tissue; in both these locations a toxic metabolite may be changed to an inactive form. Therefore, the kinetic phase involves a number of pathways by which a xenobiotic substance is converted to a toxicant that can act on a receptor or to a substance that is eliminated from the organism.

Synergism, Potentiation, and Antagonism

The biological effects of two or more toxic substances can be different in kind and degree from those of one of the substances alone.[8] One of the ways in which this can occur is when one substance affects the way in which another undergoes any of the steps in the kinetic phase as shown in Figure 3.5. Chemical interaction between substances may affect their toxicities. Both substances may act upon the same physiologic function or two substances may compete for binding to the same receptor. When both substances have the same physiologic function, their effects may be simply **additive** or they may be **synergistic** (the total effect is greater than the sum of the effects of each separately). **Potentiation** occurs when an inactive substance enhances the action of an active one and **antagonism** when an active substance decreases the effect of another active one.

3.8. BIOCHEMICAL EFFECTS OF TOXICANTS

This section briefly outlines the biochemical effects of toxic substances. Biochemical effects of toxicants can occur in any part of an organism and they constitute the basis of an organism's response to toxic substances. Therefore, biochemical interactions are very important in toxicology and they are discussed in greater detail in Chapter 4.

Following its passage through, or production in, the kinetic phase, a toxicant may undergo two major types of biochemical interactions that lead to clinical symptoms of poisoning. These are classified according to whether or not the binding of the toxicant to the receptor is reversible.[9]

Chemical Lesions

An irreversible covalent bond between a toxicant and a receptor is sometimes called a **chemical lesion**. An example of this is provided by the following reaction between sulfur-seeking mercury(II) ion and sulfhydryl (SH) groups on the active sites of enzymes:

$$\text{Enzyme}\begin{matrix}\text{SH}\\ \text{SH}\end{matrix} + Hg^{2+} \rightarrow \text{Enzyme}\begin{matrix}\text{S}\\ \text{S}\end{matrix}\text{Hg} + 2H^{+} \quad (3.1)$$

This reaction inhibits the enzyme and prevents it from performing its normal function.[10] Other heavy metals, particularly lead and cadmium, similarly inhibit enzymes. Cyanide bonds irreversibly to ferricytochrome oxidase, an iron-containing metalloenzyme crucial to the process by which the body utilizes molecular oxygen in cellular respiration. Hydrogen sulfide (H_2S) has a similar effect. Another much different example is the reaction with DNA of epoxides produced as active metabolites from microsomal processes or the reaction of DNA with biological alkylating agents; both reactions can cause carcinogenic or mutagenic effects.

Reversible Toxicant/Receptor Binding

A toxicant may react reversibly with a receptor to form a species that later dissociates to produce unaltered toxicant and receptor. In this context the toxicant is called a **ligand** and the product is a **ligand-receptor complex**. When bound in the form of a complex, a receptor does not perform its biological functions normally and a toxic effect is produced.

Major Receptors: Function and Toxic Response

As shown under "Biochemical Effect" in Figure 3.5, a living organism has several major receptors (enzymes, cell membranes) and functions (biosynthesis, metabolism) that are affected by toxicants. These are outlined briefly here.

As discussed in Chapter 2, *enzymes* are proteinaceous species that catalyze essential metabolic processes. A toxicant can impair the catalysis by binding to the enzyme as shown by the example in Equation 3.1. The toxicant reaction with the enzyme may be reversible or irreversible. Similar effects occur when the toxicant reacts with coenzymes, metal activators, or enzyme substrates.

The *cell membrane* (Section 2.7, Figure 2.11) is crucial in determining the extent to which a toxic substance penetrates cells. The membrane itself may be affected by a toxic substance because of alterations in membrane permeability which change the rate of passive (diffusion) transport of toxicants through the membrane. Carriers involved in active transport can also be disturbed, affecting this mode of cellular invasion by toxicants.

Carbohydrate metabolism and **lipid metabolism** are crucial processes that can be disrupted by toxicants. Carbohydrate metabolism involves complex sequences of steps and cycles that lead eventually to the oxidation of carbohydrates to CO_2 and H_2O with production of energy (see respiration, below). Carbohydrate biosynthesis in the body results in the production of storage glycogen. Toxicants can interfere with carbohydrate metabolism and biosynthesis. Lipids are contained in cell membranes and participate in a number of crucial processes in the cell. Excessive interference with lipid metabolism by a toxicant can result in pathological accumulations of lipids in the liver ("fatty liver").

Respiration is the overall process by which electrons are transferred to molecular oxygen in the biological oxidation of energy-yielding substrates. Part of the respiration process is the generation of adenosine triphosphate (ATP), a molecule that provides energy for synthesis and other cellular processes by virtue of two high-energy phosphate bonds in the molecule. This energy is released by the hydrolysis of ATP to yield adenosine diphosphate (ADP) via glycolysis or the citric acid cycle, both of which are enzyme-catalyzed processes. Many toxicants are known to inhibit the enzymes involved in these processes and to inhibit oxidative phosphorylation, another step in the overall respiratory process. The irreversible binding of cyanide to ferricytochrome oxidase, which prevents this crucial enzyme from performing its role in respiration, was noted earlier in this section.

Toxicants can cause **protein biosynthesis** to be stopped or to produce the wrong products. This occurs by the action of toxicants on deoxyribonucleic acid (DNA), which is contained in the cell nucleus and which "gives directions" for the synthesis of new proteins.

Metabolism is controlled by a number of **regulatory processes** that often involve hormones or enzymes. Hormones are produced by the endocrine glands, such as the pituitary gland or thyroid gland, and are transported by body fluids to various parts of the body where they regulate biosynthesis and the reverse process, catabolism. Among the aspects of hormonal activity that can be adversely affected by toxicants are the synthesis, release, and storage of hormones. As with other enzymes, toxicants can alter or totally stop the action of regulatory enzymes.

3.9. BEHAVIORAL AND PHYSIOLOGICAL RESPONSES

The final part of the overall toxicological process outlined in Figure 3.5 consists of **behavioral** and **physiological responses**, which are observable symptoms of poisoning. These are discussed here, primarily in terms of responses seen in humans and other animals. Non-animal species exhibit other kinds of symptoms from poisoning; for example, plants exhibit leaf mottling, pine needle loss, and stunted growth as a result of exposure to some toxicants.

Vital Signs

Human subjects suffering from acute poisoning usually show alterations in the **vital signs**, which consist of **temperature, pulse rate, respiratory rate**, and **blood pressure**. These are discussed here in connection with their uses as indicators of toxicant exposure.

Some toxicants that affect body temperature are shown in Figure 3.6. Among those that increase body temperature are benzadrine, cocaine, sodium fluoroacetate, tricyclic antidepressants, hexachlorobenzene, and salicylates (aspirin). In addition to phenobarbital and ethanol, toxicants that decrease body temperature include phenothiazine, clonidine, glutethimide, and haloperidol.

Toxicants may have three effects on pulse rate. These are **bradycardia** (decreased rate), **tachycardia** (increased rate), and **arrhythmia** (irregular pulse). Alcohols may cause either bradycardia or tachycardia. Amphetamines, belladonna alkaloids, cocaine, and tricyclic antidepressants may cause either tachycardia or arrhythmia.

Toxic doses of digitalis may result in bradycardia or arrhythmia. The pulse rate is decreased by toxic exposure to carbamates, organophosphates, local anesthetics, barbiturates, clonidine, muscaric mushroom toxins, and opiates. In addition to the substances mentioned above, those that cause arrhythmia are arsenic, caffeine, belladonna alkaloids, phenothizine, theophylline, and some kinds of solvents.

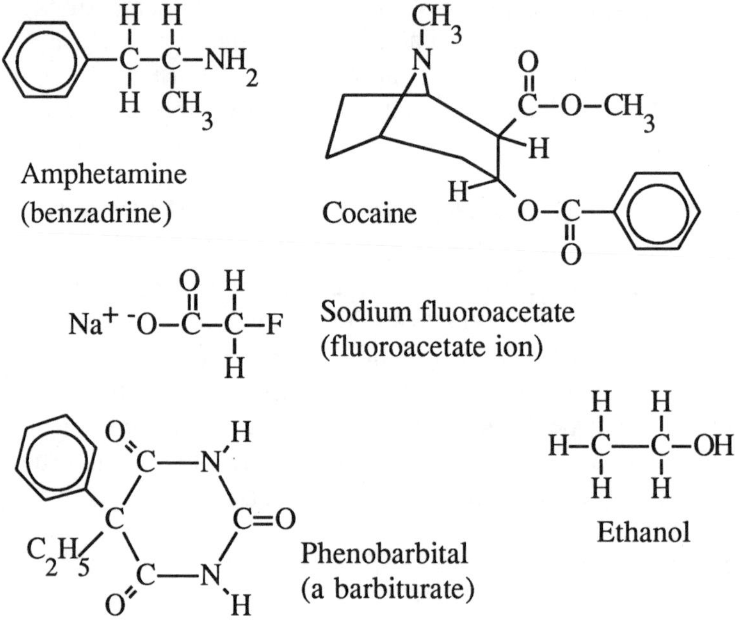

Amphetamine
(benzadrine)

Cocaine

Sodium fluoroacetate
(fluoroacetate ion)

Ethanol

Phenobarbital
(a barbiturate)

Figure 3.6. Examples of toxicants that affect body temperature. Amphetamine, cocaine, and fluoroacetate increase body temperature; phenobarbital and ethanol decrease it.

Among the toxicants that increase respiratory rate are cocaine, amphetamines, and fluoroacetate (all shown in Figure 3.6), nitrites (compounds containing the NO_2^- ion), methanol (CH_3OH), salicylates, and hexachlorobenzene. Cyanide (see respiration in Section 3.8) and carbon monoxide may either increase or decrease respiratory rate. Alcohols other than methanol, analgesics, narcotics, sedatives, phenothiazines, and opiates in toxic doses decrease respiratory rate. The structural formulas of some compounds that affect respiratory rate are shown in Figure 3.7.

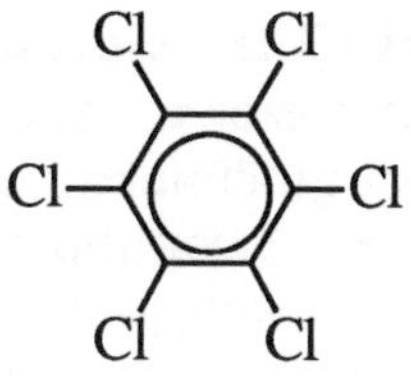

Hexachlorobenzene

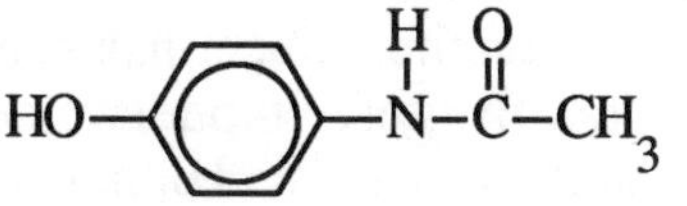

Acetaminophen

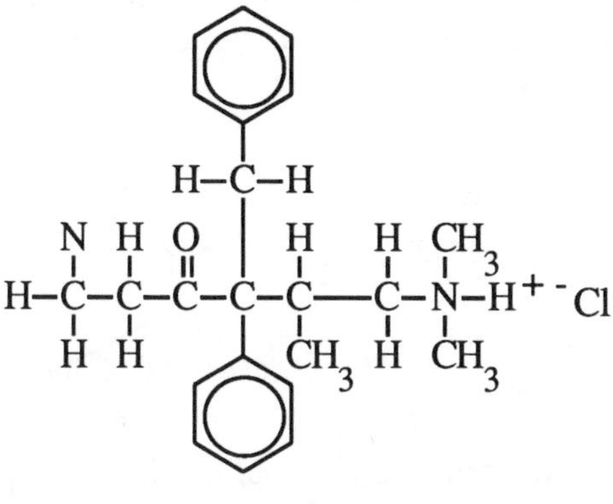

Propoxyphene hydrochloride (Darvon)

Figure 3.7. Some compounds that affect respiratory rate. Acetaminophen is one of the simple analgesics, which in therapeutic doses relieve pain without any effect upon an individual's consciousness. Propoxyphene hydrochloride is a narcotic analgesic, a class of substances which can cause biochemical changes in the body leading to chemical dependency.

Amphetamines and cocaine (see Figure 3.6), tricyclic antidepressants (see Figure 3.8), phenylcyclidines, and belladonna alkaloids at toxic levels increase blood pressure. Overdoses of antihypertensive agents decrease blood pressure, as do toxic doses of opiates, barbiturates, iron, nitrite, cyanide, and mushroom toxins.

Skin Symptoms

In many cases the skin exhibits evidence of exposure to toxic substances. The two main skin characteristics observed as evidence of poisoning are skin color and degree of skin moisture. Excessively dry skin tends to accompany poisoning by tricyclic antidepressants, antihistamines, and belladonna alkaloids. Among the toxic substances for which moist skin is a symptom of poisoning are mercury, arsenic, thallium, carbamates, and organophosphates. The skin appears flushed when the subject has been exposed to toxic doses of carbon monoxide, nitrites, amphetamines, monsodium glutamate, and tricyclic antidepressants. Higher doses of cyanide, carbon monoxide, and nitrites give the skin a **cyanotic** appearance (blue color due to oxygen deficiency in the blood). Skin may appear **jaundiced** (yellow because of the presence of bile pigments in the blood) when the subject is poisoned by a number of toxicants including arsenic, arsine gas (AsH_3), iron, aniline dyes, and carbon tetrachloride.

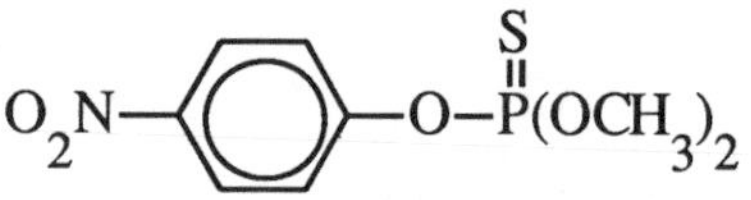

Methyl parathion

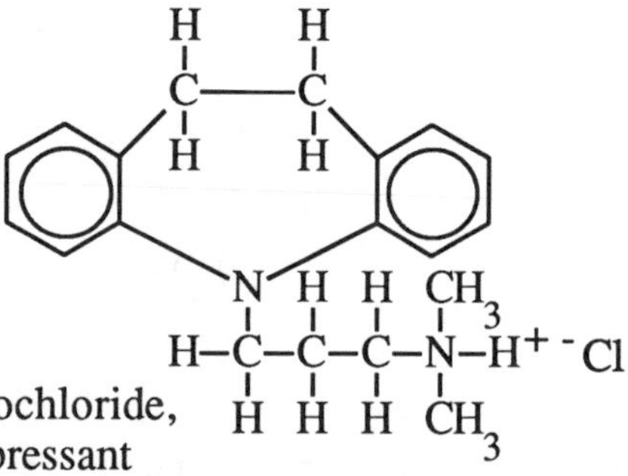

Imiprimine hydrochloride, a tricyclic antidepressant

Figure 3.8. Structures of toxicants that can affect pulse rate. Methyl parathion, a commonly used plant insecticide, can cause bradycardia. Imiprimine hydrochloride, a tricyclic antidepressant, can cause either tachycardia or arrhythmia.

Odors

Toxic levels of some materials cause the body to have unnatural **odors** because of parent compound toxicants or their metabolites secreted through the skin, exhaled through the lungs, or present in tissue samples. Some examples of odorous species are shown in Figure 3.9. In addition to the odors noted in the figure, others symptomatic of poisoning include aromatic odors from hydrocarbons and the odor of violets arising from the ingestion of turpentine. Alert pathologists have uncovered evidence of poisoning murders by noting the bitter almond odor of HCN in tissues of victims of criminal cyanide poisoning. A characteristic rotten egg odor is evidence of hydrogen sulfide (H_2S) poisoning. The same odor has been reported at autopsies of carbon disulfide poisoning victims.[11] As noted in Chapter 6, even very slight exposures to some selenium compounds cause an extremely potent garlic breath odor.

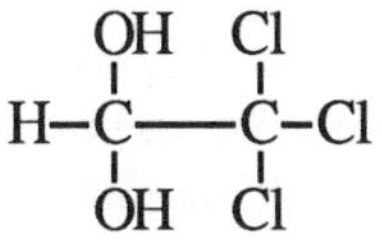

Chloral hydrate
(pear odor)

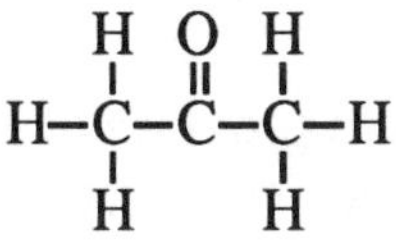

Acetone
(acetone odor)

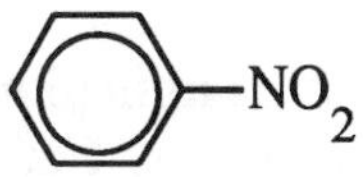

Nitrobenzene
(shoe polish)

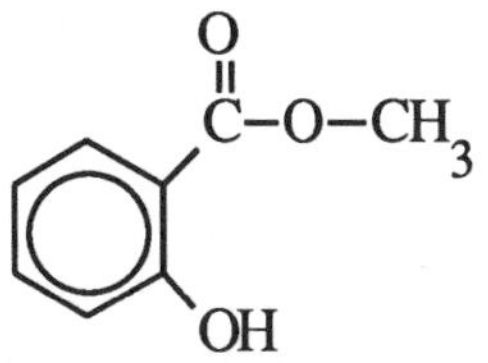

Methyl salicylate
(wintergreen)

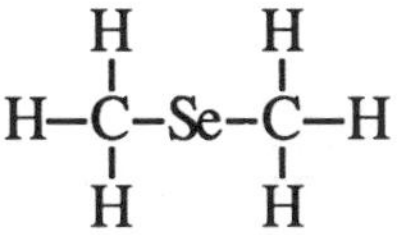

Dimethyl selenide
(garlic)

Figure 3.9. Some toxicants and the odors they produce in exposed subjects.

Eyes

Careful examination of the eyes can reveal evidence of poisoning. The response, both in size and reactivity, of the pupils to light can provide useful evidence of a response to toxicants. Both voluntary and involuntary movement of the eyes can be significant. The appearance of eye structures, including optic disc, conjunctiva, and blood vessels, can be significant. Eye **miosis**, defined as excessive or prolonged contraction of the eye pupil, is a toxic response to a number of substances, including alcohols, carbamates, organophosphates, and phenycyclidine. The opposite response **mydriasis** (excessive pupil dilation) is caused by amphetamines, belladonna alkaloids, glutethimide, and tricyclic antidepressants, among others. **Conjuctivitis** is a condition marked by inflammation of the conjunctiva, the mucus membrane that covers the front part of the eyeball and the inner lining of the eyelids. Corrosive acids and bases (alkalies) cause conjunctivitis, as do exposures to nitrogen dioxide, hydrogen sulfide, methanol, and formaldehyde. **Nystagmus**, the involuntary movement of the eyeballs, usually in a side-to-side motion, is observed in poisonings by some toxicants, including barbiturates, phenycyclidine, phentoin, and ethychlorovynol.

Mouth

Examination of the mouth provides evidence of exposure to some toxicants. Caustic acids and bases cause a moist condition of the mouth. Other toxicants that cause the mouth to be more moist than normal include mercury, arsenic, thallium, carbamates, and organophosphates. A dry mouth is symptomatic of poisoning by tricyclic antidepressants, amphetamines, antihistamines, and glutethimide.

Gastrointestinal Tract

The gastrointestinal tract responds to a number of toxic substances, usually by pain, vomiting, or paralytic ileus (see Section 3.5). Severe gastrointestinal pain is symptomatic of poisoning by arsenic or iron. Both of these substances can cause vomiting, as can acids, bases, fluorides, salicylates, and theophyllin. Paralytic ileus can result from ingestion of narcotic analgesics, tricyclic antidepressants, and clonidine.

Central Nervous System

The central nervous system responds to poisoning by exhibiting symptoms such as **convulsions**, **paralysis**, **hallucinations**, and **ataxia** (lack of coordination of voluntary movements of the body). Other behavioral symptoms of poisoning include agitation, hyperactivity, disorientation, and delirium.

Among the many toxicants that cause convulsions are chlorinated hydrocarbons, amphetamines, lead, organophosphates, and strychnine. There are several levels of **coma**, the term used to describe a lowered level of consciousness. At level 0 the subject may be awakened and will respond to questions. At level 1 withdrawal from painful stimuli is observed and all reflexes function. A subject at level 2 does not withdraw from painful stimuli, although most reflexes still function. Levels 3 and 4 are characterized by the absence of reflexes; at level 4 respiratory action is depressed and the cardiovascular system fails. Among the many toxicants that cause coma are narcotic analgesics, alcohols, organophosphates, carbamates, lead, hydrocarbons, hydrogen sulfide, benzodiazepines, tricyclic antidepressants, isoniazid, phenothiazines, and opiates.

3.10. IMMUNE SYSTEM RESPONSE

The **immune system** is the body's defense against biological systems that would harm it.[12] The most obvious of these consist of **infectious agents**, such as viruses or bacteria. Also included are **neoplastic cells**, which give rise to cancerous tissue. The immune system produces **immunoglobin**, a substance consisting of proteins bound to carbohydrates. This material functions as **antibodies** against **immunogen** or **antigen** macromolecules of polysaccharides, nucleic acids, or proteins characteristic of invasive foreign virus, bacteria, or other biological materials. The cells that the immune system uses to provide protection are called **leukocytes.**

Toxicants can adversely affect the immune system in several ways. **Immunosuppression** occurs when the body's natural defense mechanisms are impaired by agents such as toxicants. Radiation and drugs such as chemotherapeutic agents, anticonvulsants, and corticosteroids can have immunosuppressive effects. Immunosuppressants are deliberately used to prevent rejection of transplanted organs. In some cases toxicants adversely alter the mechanisms by which the immune system defends the body against pathogens and neoplastic cells. Another effect of toxicants on the immune system occurs through the loss of its ability to control proliferation of cells, resulting in leukemia or lymphoma.

Foreign agents can cause the immune system to overreact with an extreme, self-destructive response called **allergy** or **hypersensitivity**. This reaction probably occurs after the foreign agents or their metabolites become associated with large molecules endogenous to the body. Among the many substances that cause allergy are metals (beryllium, chromium, nickel), penicillin, formaldehyde, pesticides, food additives, resins, and plasticizers.

Effects upon the immune system are gaining increasing recognition as factors in toxicology and in evaluating the toxicity of various substances. There are numerous ways of evaluating potential effects upon the immune system. One is a "two-tier method" that makes use of numerous tests made directly on the test organism and on samples taken from it. The first tier consists of relatively simple tests such as measurements of body mass, blood count, examination of tissue (histology), and ability to form antibodies. The second tier consists of more sophisticated tests, such as the *Streptococcus* challenge, that measure host resistance. Bone marrow evaluations are also employed as second-tier tests.

LITERATURE CITED

1. McGuigan, Michael A., "Clinical Toxicology," Chapter 2 in *A Guide to General Toxicology*, Freddy Homburger, John A. Hayes, and Edward W. Pelikan, Eds., Karger, New York, 1983, pp. 23-69.

2. Loomis, Ted A. "Influence of Route of Administration on Toxicity," Chapter 5 in *Essentials of Toxicology*, Lea and Faebiger, Philadelphia, 1978, pp. 67-79.

3. Klaassen, Curtis D., "Distribution, Excretion and Absorption of Toxicants," Chapter 3 in *Casarett and Doull's Toxicology*, 3rd ed., Curtis D. Klaassen, Mary O. Amdur, and John Doull, Eds., Macmillan Publishing Co., New York, 1986, pp. 33-63.

4. "Percutaneous Toxicity," A. H. McCreesh, *Tox. Appl. Pharmacol.*, **20**, 1965.

5. Aldridge, W. N., "The Need to Understand Mechanisms," in *The Scientific Basis of Toxicity Assessment*, H. R. Witschi, Ed., Elsevier/North Holland Biomedical Press, Amsterdam, 1980, pp. 305-319.

6. Cohen, Gerald M., Ed., *Target Organ Toxicity*, CRC Press, Inc., Boca Raton, Florida, 1988.

7. Boyd, M. R., "Effects of Inducers and Inhibitors on Drug-Metabolizing Enzymes and on Drug Toxicity in Extrahepatic Tissues," Ciba Foundation Symposium 76, *Environmental Chemicals, Enzyme Function, and Human Disease*, Excerpta Medica Amsterdam, 1980, pp. 67-81.

8. James, Robert C., "General Principles of Toxicology," Chapter 2 in *Industrial Toxicology*, Phillip L. Williams and James L. Burson, Eds., Van Nostrand Reinhold, New York, 1985, pp. 7-26.

9. Ariens, E. J., A. M. Simonis, and J. Offermeir, *Introduction to General Toxicology*, E. J. Ariens, Ed., Academic Press, New York, 1976.

10. Manahan, Stanley E., "Environmental Biochemistry and Chemical Toxicology," Chapter 18 in *Environmental Chemistry*, Brooks/Cole Publishing Co., Monterey, CA, 1984, pp. 496-533.

11. Dreisbach, Robert H., and William O. Robertson, *Handbook of Poisoning*, 12th ed., Appleton and Lange, Norwalk, Conn., 1987.

12. Dean, Jack H., Michael L. Murray, and Edward C. Ward, "Toxic Responses of the Immune System," Chapter 9 in *Casarett and Doull's Toxicology*, 3rd ed., Curtis D. Klaassen, Mary O. Amdur, and John Doull, Eds., Macmillan Publishing Co., New York, 1986, pp. 245–285.

4

Biochemical Action and Transformation of Toxicants

4.1. BIOCHEMICAL TRANSFORMATIONS

Systemic poisons in the body undergo (1) biochemical reactions through which they have a toxic effect and (2) biochemical processes that increase or reduce their toxicities or which change toxicants to forms that are readily eliminated from the body. These processes are summarized in a general sense in Chapter 3, Figure 3.5. This chapter discusses the metabolic processes that toxicants undergo in greater detail. It places emphasis on chemical aspects and on processes that lead to products that can be eliminated from the organism.

Phase I and Phase II Reactions

The processes that most xenobiotics undergo in the body can be divided into the two categories of phase I reactions and phase II reactions.[1] A **phase I reaction** introduces reactive, polar functional groups (see Table 2.1) onto lipophilic ("fat-seeking") toxicant molecules. In their unmodified forms, such toxicant molecules tend to pass through lipid-containing cell membranes and may be bound to, and transported through the body by, lipoproteins. Because of the functional group attached, the product of a phase I reaction is usually more water-soluble than the parent xenobiotic species, and more importantly, possesses a "chemical handle" to which a substrate material in the body may become attached so that the toxicant can be eliminated from the body. The binding of such a substrate is a **phase II reaction**, and it produces a **conjugation product** that is amenable to excretion from the body.

4.2. PHASE I REACTIONS

Figure 4.1 shows the overall processes involved in a phase I reaction. Normally a phase I reaction adds a functional group to a hydrocarbon chain or ring or modifies one that is already present. The product is a chemical species that readily undergoes conjugation with some other species naturally present in the body to form a substance that can be readily excreted. Phase I reactions are discussed below.

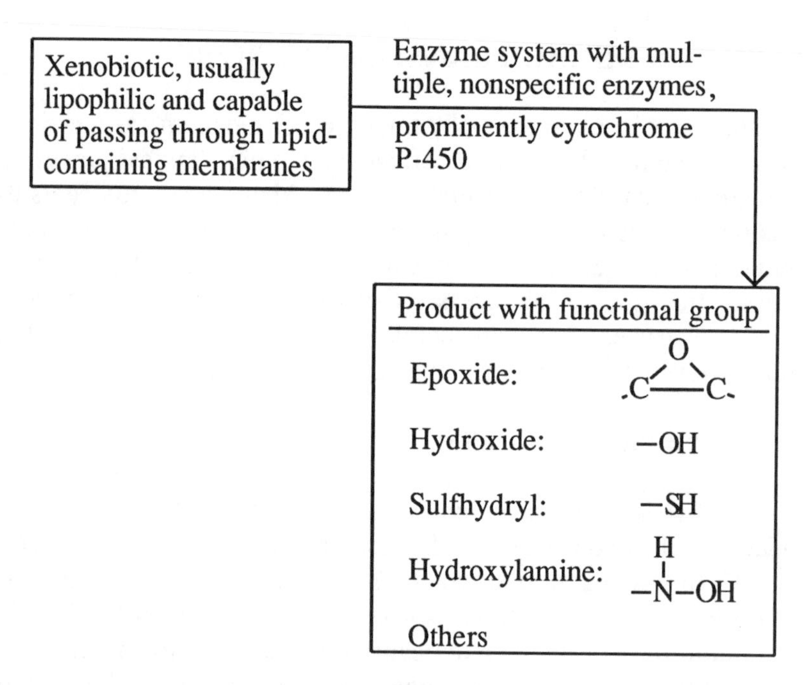

Figure 4.1. Overall process of Phase I reactions.

Mixed-Function Oxidases

The most important phase I reactions are classified as microsomal mixed-function oxidase reactions. Microsomes refer to a fraction collected from the centrifugation at about 100,000 x *g* of cell homogenates and consisting of pellets. These pellets contain rough and smooth **endoplasmic reticulum** (extensive networks of membranes in cells) and Golgi bodies, which store newly synthesized molecules.[2] **Mixed-function oxidations** occur with O_2 as the oxidizing agent, one atom of which is incorporated into the substrate, and the other going to form water:

$$\text{Substrate} + O_2 \xrightarrow[\text{oxidation}]{\text{Mixed-function}} \begin{cases} \text{Product-OH} \\ H_2O \end{cases} \quad (4.1)$$

The key enzyme of the system is cytochrome P-450, the active site of which contains an iron atom that can change between the +2 and +3 oxidation states.[3] The enzyme can bind to the substrate and molecular O_2 as part of the process by which the substrate is oxidized. Cytochrome P-450 is found most abundantly in the livers of vertebrates, reflecting the liver's role as the body's primary defender against systemic poisons. Cytochrome P-450 occurs in many other parts of the body, such as the kidney, ovaries, testes, and blood. The presence of this enzyme in the lung, skin, and gastrointestinal tract may reflect their defensive roles against toxicants.

Epoxidation consists of adding an oxygen atom between two C atoms in an unsaturated system as shown in Figures 4.2 and 4.3. It is

$$\underset{\text{Trichloroethylene}}{Cl_2C{=}CHCl} \xrightarrow[\text{epoxidation}]{O_2\text{, enzyme-mediated}} \text{Cl–C(Cl)–O–C(H)–Cl (epoxide)} \rightarrow \underset{\text{Trichloroacetaldehyde}}{Cl_3C{-}CHO} \quad (4.2)$$

$$\text{Benzene} \xrightarrow[\text{epoxidation}]{O_2\text{, enzyme-mediated}} \text{Benzene epoxide} \quad (4.3)$$

a particularly important means of metabolic attack upon aromatic rings that abound in many xenobiotic compounds. Cytochrome P-450 is involved in epoxidation reactions. Both of the epoxidation reactions shown above have the effect of increasing the toxicities of the parent compounds, a process called **intoxication.** Some epoxides are unstable, tending to undergo further reactions, usually hydroxylation (see below). A well-known example of the formation of a stable epoxide is the conversion to aldrin of the insecticide dieldrin discussed in Chapter 11.

Hydroxylation

Hydroxylation is the attachment of –OH groups to hydrocarbon chains or rings. It can follow epoxidation as shown by the following rearrangement reaction for benzene epoxide:

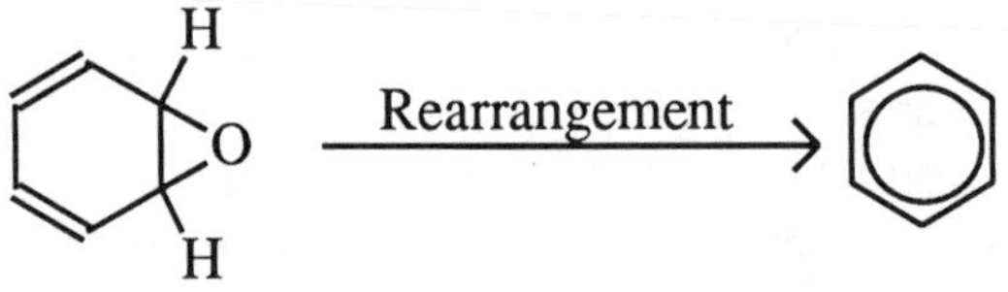

H
O
H
Rearrangement

Hydroxylation can consist of the addition of more than one epoxide group. Hydroxylation and epoxidation are responsible for making several xenobiotic compounds toxic through metabolic processes. A prominent example of this phenomenon is the metabolic production of the carcinogenic 7,8-diol-9,10-epoxide of benzo(a)pyrene as illustrated by the overall reaction shown in Figure 4.2. Benzo(a)pyrene is classified as a procarcinogen[4], or precarcinogen, in that metabolic action is required to convert it to a species that is carcinogenic per se (see later discussion).

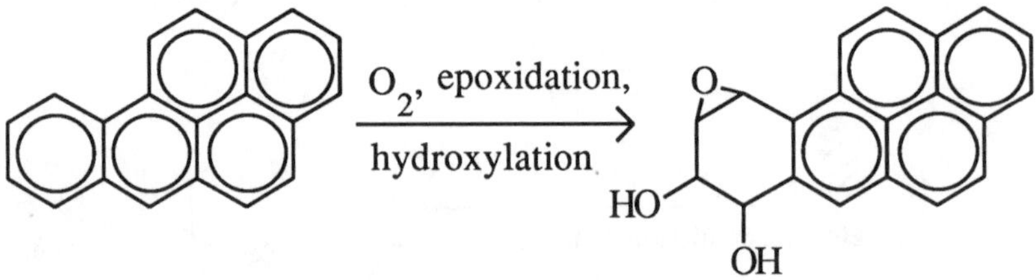

O_2, epoxidation,
hydroxylation
O
HO
OH

Figure 4.2 Epoxidation/hydroxylation of benzo(a)pyrene (left) to form carcinogenic 7,8-diol-9,10-benzo(a)pyrene epoxide (right).

Oxidation of Non-Carbon Elements

As summarized in Figure 4.3, the oxidation of nitrogen, sulfur, and phosphorus is an important type of metabolic reaction in xenobiotic compounds. It can be an important intoxication mechanism.

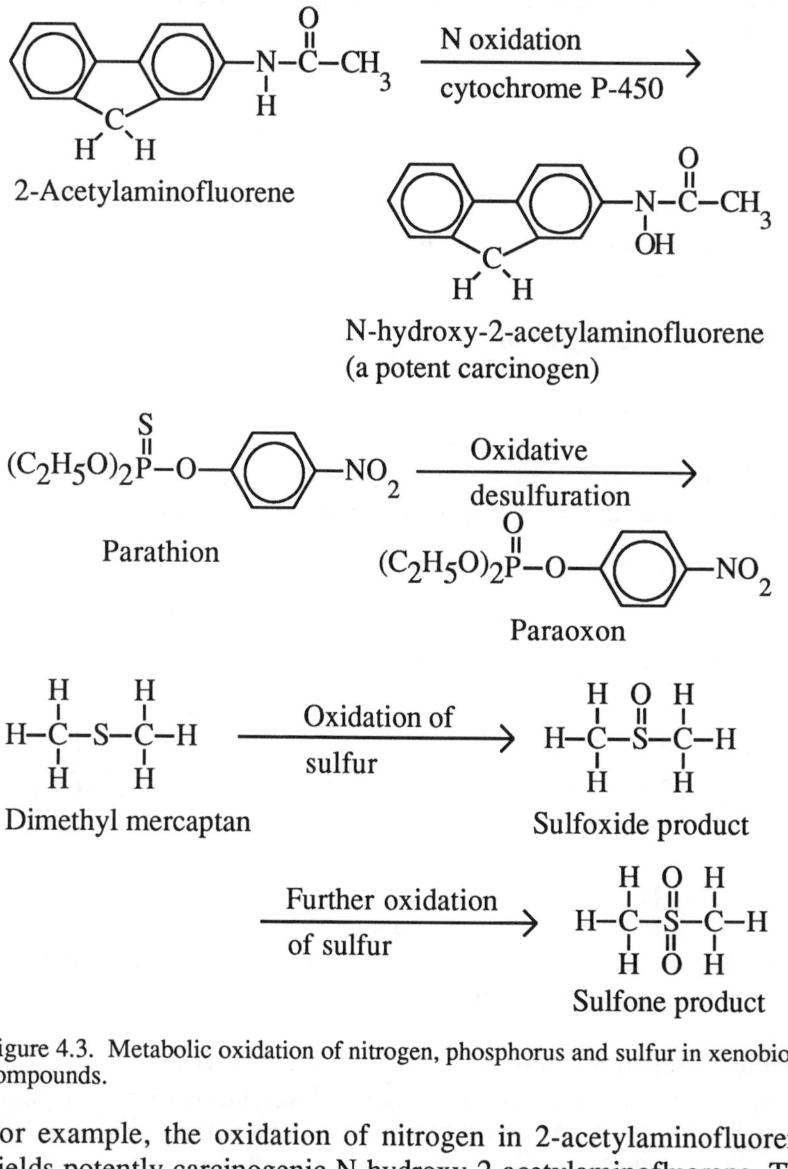

N oxidation

cytochrome P-450

2-Acetylaminofluorene

N-hydroxy-2-acetylaminofluorene
(a potent carcinogen)

$(C_2H_5O)_2P(=S)-O-C_6H_4-NO_2$

Oxidative desulfuration

Parathion

$(C_2H_5O)_2P(=O)-O-C_6H_4-NO_2$

Paraoxon

$H_3C-S-CH_3$

Oxidation of sulfur

$H_3C-S(=O)-CH_3$

Dimethyl mercaptan

Sulfoxide product

Further oxidation of sulfur

$H_3C-S(=O)_2-CH_3$

Sulfone product

Figure 4.3. Metabolic oxidation of nitrogen, phosphorus and sulfur in xenobiotic compounds.

For example, the oxidation of nitrogen in 2-acetylaminofluorene yields potently carcinogenic N-hydroxy-2-acetylaminofluorene. The oxidation of phosphorus in parathion (replacement of S by O, oxidative desulfurization) yields insecticidal paraoxon, which is much more effective than the parent compound in inhibiting acetylcholinesterase enzyme (see Section 1.10, Reaction 1.1). Two major steps in the metabolism of the plant systemic insecticide aldicarb (Figure 4.4) are oxidation to the sulfoxide followed by oxidation to the sulfone (see discussion of sulfur compounds in Chapter 12).

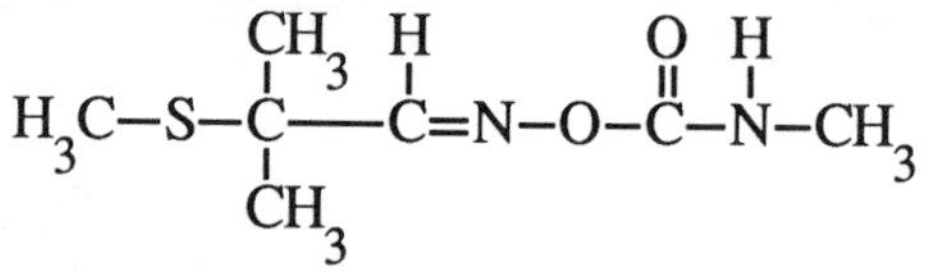

$$H_3C-S-C(CH_3)_2-C(H){=}N-O-C(=O)-N(H)-CH_3$$

Figure 4.4. Structure of the plant systemic insecticide temik (aldicarb). The sulfur is metabolically oxidizable.

Metabolic Dealkylation

Many xenobiotics contain alkyl groups, such as the methyl ($-CH_3$) group, attached to atoms of O, N, and S. An important step in the metabolism of many of these compounds is replacement of alkyl groups by H as shown in Figure 4.5.[5] These reactions are carried out

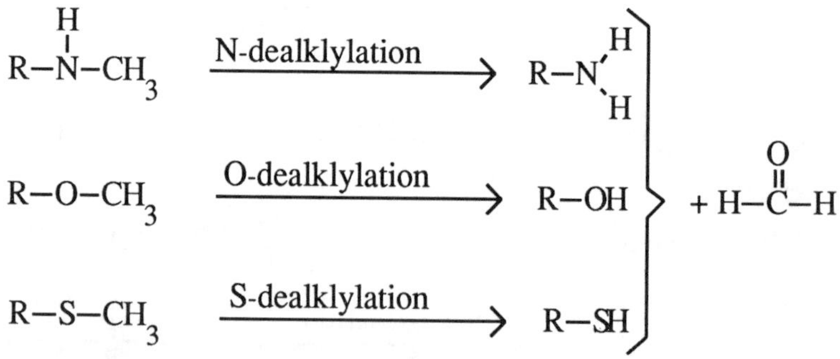

$$R-N(H)-CH_3 \xrightarrow{\text{N-dealklylation}} R-NH_2$$

$$R-O-CH_3 \xrightarrow{\text{O-dealklylation}} R-OH \quad + \; H-C(=O)-H$$

$$R-S-CH_3 \xrightarrow{\text{S-dealklylation}} R-SH$$

Figure 4.5. Metabolic dealkylation reactions shown for the removal of CH_3 from N, O and S atoms in organic compounds.

by mixed-function oxidase enzyme systems. Examples of these kinds of reactions with xenobiotics include O-dealkylation of methoxychlor insecticides, N-dealkylation of carbaryl insecticide, and S-dealkylation of dimethyl mercaptan.

Metabolic Reductions

Table 4.1 summarizes some of the major functional groups in xenobiotics that can be reduced metabolically. Reductions are carried out by **reductase enzymes**; for example, nitroreductase enzyme catalyzes the reduction of the nitro group. Reductase enzymes are found largely in the liver and to a certain extent in other organs, such as the kidney and lung. They also occur in intestinal bacteria and reduction of xenobiotics can occur in the intestinal tract.

Table 4.1. Functional Groups that Undergo Metabolic Reduction

Functional group	Process	Product
$R-C(=O)-H$	Aldehyde reduction	$R-CH_2-OH$
$R-C(=O)-R'$	Ketone reduction	$R-CH(OH)-R'$
$R-S(=O)-R'$	Sulfoxide reduction	$R-S-R'$
$R-SS-R'$	Disulfide reduction	$R-SS-H$
$H_2C=CH_2$	Alkene reduction	H_3C-CH_2-OH
$R-NO_2$	Nitro reduction	$R-NO$, $R-NH_2$, $R-NHOH$
As(V)	Arsenic reduction	As(III)

Metabolic Hydrolysis Reactions

Hydrolysis involves the addition of H2O to a molecule accompanied by cleavage of the molecule into two species. The two most common types of the compounds that undergo hydrolysis are esters and amides (see Chapters 9 and 10) as shown by the generalized reaction

$$R{-}\overset{\displaystyle O}{\overset{\|}{C}}{-}X\begin{matrix}\diagup R' \\ \diagdown (R'')\end{matrix} + H_2O \rightarrow R{-}\overset{\displaystyle O}{\overset{\|}{C}}{-}OH + HX\begin{matrix}\diagup R' \\ \diagdown (R'')\end{matrix} \quad (4.4)$$

where X is an O (in an ester) or N (in an amide, to which R' *and* R" are bonded). Organophosphate esters (see Chapter 13) also undergo hydrolysis as shown below for the plant systemic insecticide demeton:

$$(C_2H_5O)_2\overset{\displaystyle O}{\overset{\|}{P}}{-}S{-}\underset{H}{\overset{H}{C}}{-}\underset{H}{\overset{H}{C}}{-}S{-}\underset{H}{\overset{H}{C}}{-}\underset{H}{\overset{H}{C}}{-}H + H_2O \rightarrow$$

$$(C_2H_5O)_2\overset{\displaystyle O}{\overset{\|}{P}}{-}OH + HS{-}\underset{H}{\overset{H}{C}}{-}\underset{H}{\overset{H}{C}}{-}S{-}\underset{H}{\overset{H}{C}}{-}\underset{H}{\overset{H}{C}}{-}H \quad (4.5)$$

Many xenobiotic compounds, such as pesticides, are esters, amides, or organophosphate esters, and hydrolysis is a very important aspect of their metabolic fates. The types of enzymes that bring about hydrolysis are **hydrolase enzymes.** Like most enzymes involved in the metabolism of xenobiotic compounds, hydrolase enzymes occur prominently in the liver. They also occur in tissue lining the intestines, nervous tissue, blood plasma, the kidney, and muscle tissue. Enzymes that enable the hydrolysis of esters are called **esterases**, and those that hydrolyze amides are **amidases**. Aromatic esters are hydrolyzed by the action of aryl esterases and alkyl esters by aliphatic esterases.

Removal of Halogen

An important step in the metabolism of the many xenobiotic compounds that contain covalently bound halogens (F, Cl, Br, I) is the removal of halogen atoms, a process called **dehalogenation.** This may occur by **reductive dehalogenation** in which the halogen atom is replaced by hydrogen, or two atoms are lost from adjacent carbon atoms, leaving a carbon-carbon double bond. These processes are illustrated by the two following reactions:

$$\boxed{\text{Xenobiotic molecule}}\!>\!\underset{\quad}{\overset{\text{H}}{\underset{|}{\overset{|}{\text{C}}}}}-\underset{\text{Cl}}{\overset{\text{Cl}}{\underset{|}{\overset{|}{\text{C}}}}}-\text{Cl} \rightarrow \boxed{\text{Dehalogenated xenobiotic molecule}}\!>\!\overset{\text{H}}{\overset{|}{\text{C}}}-\underset{\text{Cl}}{\overset{\text{H}}{\underset{|}{\overset{|}{\text{C}}}}}-\text{Cl} \quad (4.6)$$

$$\boxed{\text{Xenobiotic molecule}}\!>\!\overset{\text{H}}{\overset{|}{\text{C}}}-\underset{\text{Cl}}{\overset{\text{Cl}}{\underset{|}{\overset{|}{\text{C}}}}}-\text{Cl} \rightarrow \boxed{\text{Dehalogenated xenobiotic molecule}}-\overset{\text{H}}{\overset{|}{\text{C}}}=\text{C}\begin{matrix}\text{Cl}\\ \text{Cl}\end{matrix} \quad (4.7)$$

Oxidative dehalogenation occurs when oxygen is added in place of a halogen atom as shown below:

$$\boxed{\text{Xenobiotic molecule}}\!>\!\overset{\text{H}}{\overset{|}{\text{C}}}-\underset{\text{Cl}}{\overset{\text{Cl}}{\underset{|}{\overset{|}{\text{C}}}}}-\text{Cl} \xrightarrow{O_2} \boxed{\text{Dehalogenated xenobiotic molecule}}\!>\!\overset{\text{H}}{\overset{|}{\text{C}}}-\overset{\text{O}}{\overset{\|}{\text{C}}}-\text{OH} \quad (4.8)$$

4.3. PHASE II REACTIONS OF TOXICANTS

Phase II reactions are also known as **conjugation reactions** because they involve the joining together of a substrate compound with another species that occurs normally in (is endogenous to) the body. This can occur with unmodified xenobiotic compounds, xenobiotic compounds that have undergone phase I reactions, and non-xenobiotic compounds. The substance that binds to these species is called a **conjugating agent.** The overall process for the conjugation of a xenobiotic compound is shown in Figure 4.6. Such a compound contains functional groups, often added as the consequence of a phase I reaction, that serve as "chemical handles" for the attachment of the conjugating agent. The product is usually less lipid-soluble, more soluble in water, less toxic, and more easily eliminated than the parent compound.

Conjugation by Glucuronides

Glucuronides are the most common endogenous conjugating agents in the body. They react with xenobiotics through the action of

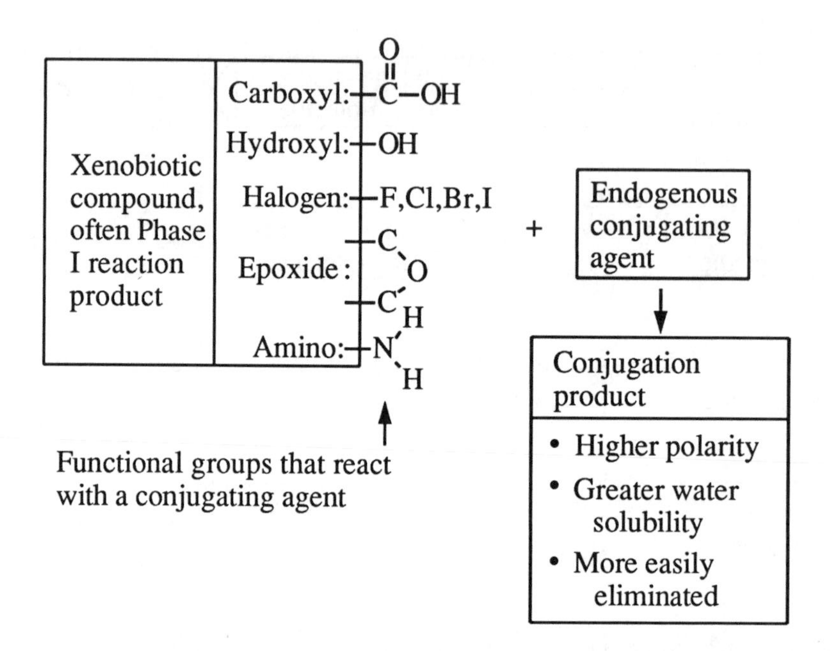

uridine diphosphate glucose (UDPG) and uridine diphosphate glucuronic acid (UDPGA). The uridinediphosphate (UDP) portion of these structures is moderately complicated and is not given here. A generalized conjugation reaction of UDPGA with a xenobiotic compound can be represented as the following:

UDGPA –O–UDP + HX–R (Xenobiotic) → –X–R (Conjugate of xenobiotic with glucuronide) + UDP (4.9)

In this reaction HX-R represents the xenobiotic species in which HX is a functional group (such as –OH) and R is an organic moiety, such as the phenyl group (benzene ring less a hydrogen atom). The kind of enzyme that mediates this type of reaction is UDP glucuronyltransferase.

Glucuronide conjugation products may be classified according to the element to which the glucuronide is bound. These elements are oxygen, nitrogen, and sulfur, and example glucuronides involving O, N, and S atoms are shown in Figure 4.7. When the functional group

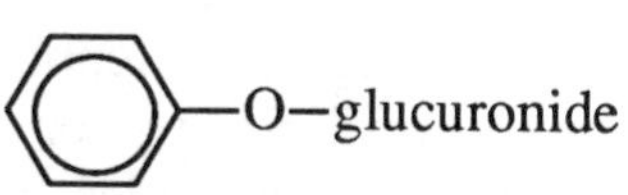

Phenylglucuronide, an O-glucuronide

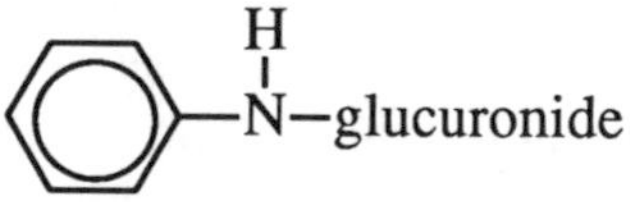

Aniline glucuronide, an N-glucuronide

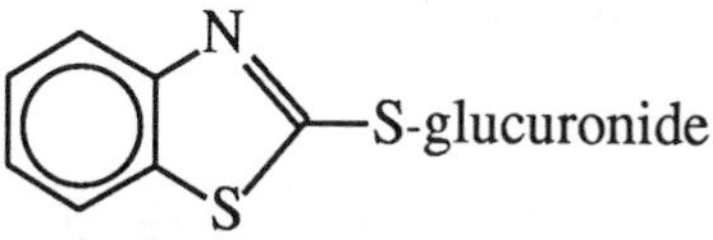

2-Mercaptothiazole-S-glucuronide, an S-glucuronide

Figure 4.7. Examples of O-, N-, and S-glucuronides.

through which conjugation occurs is a hydroxyl group, –OH (HX– in Reaction 4.9) an ether glucuronide is formed. A carboxylic acid group for HX gives an ester glucuronide. Glucuronides may be attached directly to N as the linking atom, as is the case with aniline glucuronide in Figure 4.7, or through an intermediate O atom. An example of the latter is N-hydroxyacetylaminoglucuronide, for which the structure is shown in Figure 4.8. This species is of interest because it is a stronger carcinogen than its parent xenobiotic compound, N-hydroxyacetylaminofluorene, contrary to the decrease in toxicity that usually results from glucuronide conjugation.

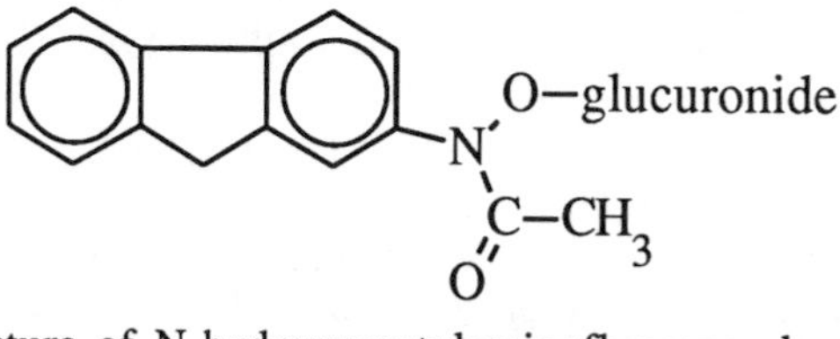

Figure 4.8. Structure of N-hydroxyacetylaminofluorene glucuronide, a more potent carcinogen than its parent compound, N-hydroxyacetylaminofluorene.

Conjugation by Glutathione

Glutathione is a crucial conjugating agent in the body. This compound is a tripeptide, meaning that it is composed of three amino acids linked together. These amino acids and their abbreviations are glutamic acid (Glu), cysteine (Cys), and glycine (Gly, structural formula in Figure 2.11). The formula of glutathione may be represented as illustrated in Figure 4.9, where the SH is shown specifically because of its crucial role in forming the covalent link to a xenobiotic compound. Glutathione conjugate may be excreted directly, or after further biochemical reactions that produce mercapturic acids (compounds with N-acetylcystein attached). The overall process just outlined as applied to a generic xenobiotic species, HX-R (see previous discussion) is illustrated in Figure 4.9.

Glu—Cys(SH)—Gly (Glutathione) + HX-R (Xenobiotic) —Glutathione transferase→ Glu—Cys(S—X—R)—Gly (Glutathione conjugate)

Glutathione conjugate → Direct excretion in bile

Loss of glutamyl and glycinyl ↓

R-X-S—CH₂—CH(NH₂)—C(=O)—OH (Cysteine conjugate) → Direct excretion in bile

Acetylation (addition of —C(=O)—CH₃) ↓

R-X-S—CH₂—CH(NH—C(=O)—CH₃)—C(=O)—OH Readily excreted mercapturic acid conjugate

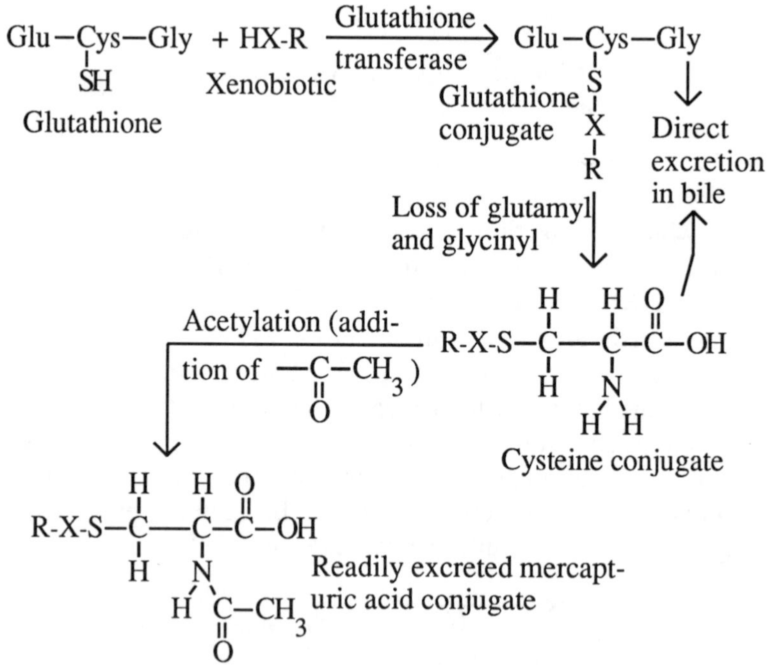

Figure 4.9. Glutathione conjugate of a xenobiotic species (HX-R) followed by formation of glutathione and cysteine conjugate intermediates (which may be excreted in bile) and acetylation to form readily excreted mercapturic acid conjugate.

There are numerous variations on the general mechanism outlined in Figure 4.9. Glutathione forms conjugates with a wide variety of xenobiotic species including alkenes, alkyl epoxides (1,2-epoxyethylbenzene), arylepoxides (1,2-epoxynaphthalene), aromatic hydrocarbons, aryl halides, alkyl halides (methyl iodide), and aromatic nitro compounds. The glutathione transferase enzymes required for the initial conjugation are widespread in the body.

Conjugation by Sulfate

Although conjugation by sulfate requires the input of substantial amounts of energy, it is very efficient in eliminating xenobiotic species through urine because the sulfate conjugates are completely ionized and therefore highly water-soluble. The major types of species that form sulfate conjugates are alcohols, phenols, and arylamines as shown by the examples in Figure 4.10. The name of the enzyme that catalyzes these reactions is sulfotransferase. The sulfating agent is a rather complex biomolecule called adenosine 3'-phosphate-5'-phosphosulfate, commonly abbreviated PAPS.

Miscellaneous Phase II Reactions

In addition to the important phase II reactions covered in the preceding sections, several other reactions of this type should be mentioned here. **Acetylation reactions** catalyzed by acetyltransferase enzymes involve the attachment of the acetyl moiety, shown as a final step in glutathione conjugation and the production of a mercapturic acid conjugate in Figure 4.9. Amino acids, particularly glycine, form conjugates with a number of xenobiotic compounds to give **peptide conjugates** that can be excreted from the body. **Methylation** of xenobiotics occurs with attachment of the $-CH_3$ group (see methylation reactions in Chapter 6).

4.4. INTERFERENCE WITH ENZYME ACTION

Enzymes are extremely important because they must function properly to enable essential metabolic processes to occur in cells. Substances that interfere with the proper action of enzymes obviously

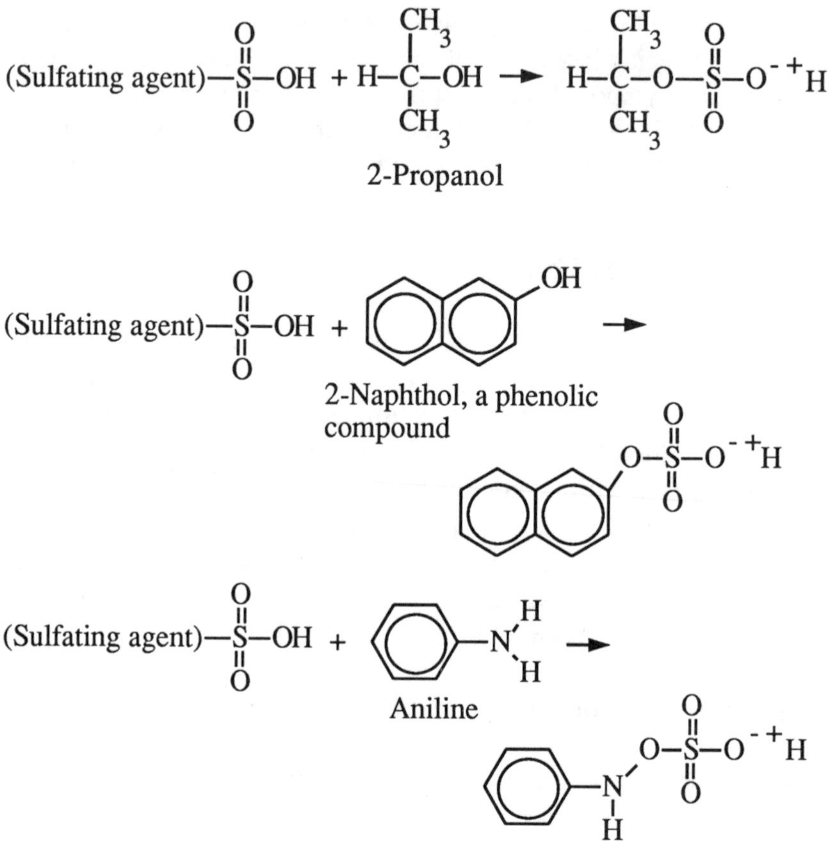

$$\text{(Sulfating agent)}-\overset{\text{O}}{\underset{\text{O}}{\overset{\|}{\underset{\|}{\text{S}}}}}-\text{OH} + \text{H}-\overset{\text{CH}_3}{\underset{\text{CH}_3}{\overset{|}{\underset{|}{\text{C}}}}}-\text{OH} \rightarrow \text{H}-\overset{\text{CH}_3}{\underset{\text{CH}_3}{\overset{|}{\underset{|}{\text{C}}}}}-\text{O}-\overset{\text{O}}{\underset{\text{O}}{\overset{\|}{\underset{\|}{\text{S}}}}}-\text{O}^{-}\,{}^{+}\text{H}$$

2-Propanol

$$\text{(Sulfating agent)}-\overset{\text{O}}{\underset{\text{O}}{\overset{\|}{\underset{\|}{\text{S}}}}}-\text{OH} + \text{(2-naphthyl)}-\text{OH} \rightarrow \text{(2-naphthyl)}-\text{O}-\overset{\text{O}}{\underset{\text{O}}{\overset{\|}{\underset{\|}{\text{S}}}}}-\text{O}^{-}\,{}^{+}\text{H}$$

2-Naphthol, a phenolic compound

$$\text{(Sulfating agent)}-\overset{\text{O}}{\underset{\text{O}}{\overset{\|}{\underset{\|}{\text{S}}}}}-\text{OH} + \text{C}_6\text{H}_5-\text{NH}_2 \rightarrow \text{C}_6\text{H}_5-\underset{\text{H}}{\underset{|}{\text{N}}}-\text{O}-\overset{\text{O}}{\underset{\text{O}}{\overset{\|}{\underset{\|}{\text{S}}}}}-\text{O}^{-}\,{}^{+}\text{H}$$

Aniline

Figure 4.10. Formation of sulfate conjugates with xenobiotic compounds.

have the potential to be toxic. Many xenobiotics that adversely affect enzymes are **enzyme inhibitors** that slow down or stop enzymes from performing their normal functions as biochemical catalysts. Stimulation of the body to make enzymes that serve particular purposes, a process called **enzyme induction**, is also important in toxicology.

The body contains numerous endogenous enzyme inhibitors that serve to control enzyme-catalyzed processes. When a toxicant acts as an enzyme inhibitor, however, an adverse effect usually results. An important example of this is the action of ions of heavy metals, such as mercury (Hg^{2+}), lead (Pb^{2+}), and cadmium (Cd^{2+}), which have strong tendencies to bind to sulfur-containing functional groups, especially –SS–, –SH, and $-S-CH_3$. These functional groups are often present on the active sites of enzymes, which, because of their specific three-dimensional structures, bind with high selectivity to the substrate species upon which the enzymes act. Toxic metal ions may bind strongly to sulfur-containing functional groups in enzyme active sites, thereby inhibiting the action of the enzyme. Such a reaction is illustrated in Figure 4.11 for Hg^{2+} ion binding to sulfhydryl groups on an enzyme active site:[6]

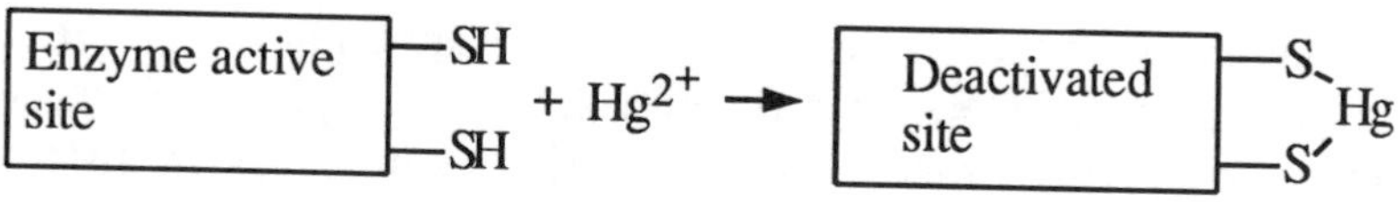

Enzyme active site	–SH –SH	+ Hg^{2+} →	Deactivated site	–S–Hg–S–

Figure 4.11. Binding of a heavy metal to an enzyme active site.

Inhibition of Metalloenzymes

Substitution of foreign metals for the metals in metalloenzymes (those that contain metals as part of their structures) is an important mode of toxic action by metals. A common mechanism for cadmium toxicity is the substitution of this metal for zinc, a metal that is present in many metalloenzymes. This substitution occurs readily because of the chemical similarities between the two metals (for example, Cd^{2+} and Zn^{2+} behave very much alike in solution). Some enzymes that are affected adversely by the substitution of cadmium for zinc are adenosine triphosphate, alcohol dehydrogenase, and carbonic anhydrase.

Inhibition by Organic Compounds

The covalent bonding of organic xenobiotic compounds to enzymes as shown in Equation 4.10 can cause enzyme inhibition. Such

$$(C_3H_7O)_2\overset{\overset{O}{\|}}{P}\text{—F} + \text{HO—(acetylcholinesterase)} \rightarrow$$

$$\text{HF} + (C_3H_7O)_2\overset{\overset{O}{\|}}{P}\text{—O—(acetylcholinesterase)} \quad (4.10)$$

bonding occurs most commonly through hydroxyl (–OH) groups on enzyme active sites. Covalent bonding of xenobiotic compounds is one of the major ways in which acetylcholinesterase (an enzyme crucial to the function of nerve impulses, see Section 1.10) can be inhibited. An organophosphate compound, such as the nerve gas compound diisopropylphosphorfluoridate (a reactant in Equation 4.10), may bind to acetylcholinesterase, thereby inhibiting the enzyme.

4.5. BIOCHEMISTRY OF MUTAGENESIS

Mutagenesis is the phenomenon in which inheritable traits result from alterations of DNA (see Section 2.7). Although mutation is a normally occurring process that gives rise to diversity in species, most mutations are harmful. The toxicants that cause mutations are known as **mutagens**. These toxicants, often the same as those that cause cancer or birth defects, are a major toxicological concern.

To understand the biochemistry of mutagenesis, it is important to know that DNA contains the nitrogenous bases adenine, guanine, cytosine, and thymine. The order in which these bases occur in DNA determines the nature and structure of newly-produced RNA, a substance produced as a step in the synthesis of new proteins and enzymes in cells. Exchange, addition, or deletion of any of the nitrogenous bases in DNA alters the nature of RNA produced and can change vital life processes, such as the synthesis of an important enzyme.[7] This phenomenon, which can be caused by xenobiotic compounds, is a mutation that can be passed on to progeny, usually with detrimental results.

There are several ways in which xenobiotic species may cause mutations. It is beyond the scope of this work to discuss these mechanisms in detail. For the most part, however, mutations due to xenobiotic substances are the result of chemical alterations of DNA, such as those discussed in the example below.

Nitrous acid, HNO_2, is an example of a chemical mutagen that is often used to cause mutations in bacteria. To understand the mutagenic activity of nitrous acid it should be noted that three of the nitrogenous bases — adenine, guanine, and cytosine — contain the amino group, $-NH_2$. The action of nitrous acid is to replace amino groups with doubly bonded oxygen atoms, thereby placing keto groups (C=O) in the rings of the nitrogenous bases and converting them to other compounds. When this occurs, the DNA may not function in the intended manner, causing a mutation to occur.

Tris

One of the more widely publicized mutagens is tris(2,3-dibromopropyl)phosphate, commonly called "tris," that was used as a flame retardant in children's sleepwear.[8] Tris was found to be mutagenic in experimental animals and metabolites of it were found in children wearing the treated sleepwear. This strongly suggested that tris is absorbed through the skin and its uses were discontinued.

4.6. BIOCHEMISTRY OF CARCINOGENESIS

Cancer is a condition characterized by the uncontrolled replication and growth of the body's own cells (somatic cells). It is now generally believed that many — and perhaps most — cancers are started by the action of synthetic and naturally occurring chemicals (in some cases viruses cause cancer). The role of xenobiotic chemicals in causing cancer is called **chemical carcinogenesis**. It is often regarded as the single most important facet of toxicology and clearly the one that receives the most publicity.

Despite large expenditures of time and money on the subject, the biochemical bases of chemical carcinogenesis are not well understood. The overall processes for the induction of cancer may be quite complex, involving numerous steps. It is generally recognized that there are two major steps: an initiation stage followed by a promotional stage.[9] Chemical carcinogens are often mutagens and it is believed that in many cases cancerous cells result from mutations of normal cells exposed to carcinogens. This implies that chemical carcinogens alter DNA in a manner such that an "outlaw cell" is formed that continues to replicate itself and form cancerous tissue.

Alkylating Agents in Carcinogenesis

Chemical carcinogens usually have the ability to form covalent bonds with macromolecular life molecules, especially DNA.[10] Prominent among these are the **alkylating agents** which attach alkyl groups — such as methyl (CH_3) or ethyl (C_2H_5) — to DNA. A similar type of compound, **arylating agents**, act to attach aryl

moieties, such as the phenyl group

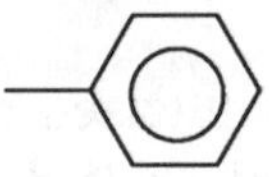

to DNA. The alkyl and aryl groups become attached to N and O atoms in the nitrogenous bases that compose DNA; examples are shown in Figure 4.12. This leads to alteration in the DNA and it can result in the growth and replication of neoplastic (cancerous) cells.

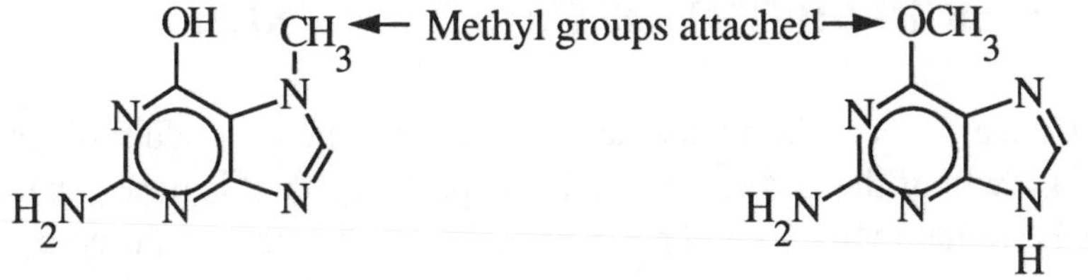

Figure 4.12. Alkylated (methylated) forms of the nitrogenous base guanine.

Primary Carcinogens and Procarcinogens

In order for them to cause cancer, most cancer-causing substances require metabolic activation and are called **precarcinogens** or **procarcinogens**. The metabolic species actually responsible for carcinogenesis, usually by its interaction with DNA, is termed an **ultimate carcinogen**. Some species that are intermediate metabolites between precarcinogens and ultimate carcinogens are called **proximate carcinogens.**[11] Carcinogens that do not require biochemical activation are categorized as **primary** or **direct-acting carcinogens**.

Testing for Carcinogens

In some cases chemicals are known to be carcinogens from epidemiological studies of exposed humans. Animals are used to test for carcinogenicity, and the results can be extrapolated with some uncertainty to humans. The most broadly applicable test for potential carcinogens is the **Bruce Ames** procedure, which actually reveals mutagenicity. The principle of this method is the reversion of mutant histidine-requiring *Salmonella* bacteria back to a form that can synthesize their own histidine.[12] The bacteria are inoculated onto a medium that does not contain histidine, and those that mutate back to a form that can synthesize histidine establish colonies which are assayed on the growth medium, thereby providing both a qualitative and quantitative indication of mutagenicity. The test chemicals are mixed with homogenized liver tissue to simulate the body's alteration of chemicals (conversion of procarcinogens to ultimate carcinogens). Up to 90% correlation has been found between mutagenesis on this test and known carcinogenicity of test chemicals.

4.7. BIOCHEMISTRY OF TERATOGENESIS

Teratology is the science of birth defects caused by radiation, viruses, and chemicals, including drugs.[13] Xenobiotic chemical species that cause birth defects are called **teratogens**. Teratogens affect developing embryos adversely, often with remarkable specificity in regard to effect and stage of embryo development when exposed. A teratogen may cause a specific effect when exposure occurs on a definite number of days after conception; if exposure occurs only a few days sooner or later, no effect, or an entirely different one, may be observed. Although mutations in germ cells (egg or sperm cells) may cause birth defects (e.g., Down's syndrome), teratology usually deals with defects arising from damage to embryonic or fetal cells.

The biochemical effects of teratogens are varied and, for the most part, not well understood. In some cases teratogens interfere with DNA synthesis. Teratogens may alter the function of nucleic acids in cell replications, and adverse effects may result. Serious defects may arise from either an absence or excess of chromosomes caused by exposure to xenobiotics, an effect that sometimes can be revealed by microscopic examination. Enzyme inhibition (see Section 4.4) by xenobiotics can be teratogenic. Xenobiotics that deprive the fetus of essential substrates (for example, vitamins), that interfere with energy supply, or that alter the permeability of the placental membrane may all cause birth defects.

Thalidomide

Perhaps the most notorious teratogen is thalidomide (see below), a sedative-hypnotic drug used in Europe and Japan in 1960-1961. Some infants born to women who had taken thalidomide from days 35 through 50 of their pregnancies were born suffering from amelia or phocomelia, the absence or severe shortening, respectively, of the limbs. About 10,000 children were affected.

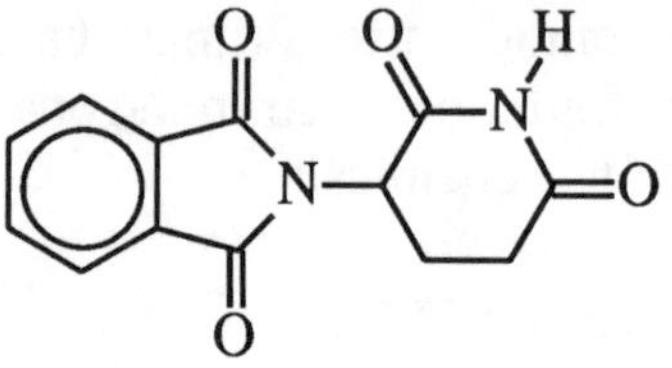

Thalidomide

Accutane

In 1988 the U.S. Food and Drug Administration estimated that **accutane** (retinoic acid) used as an anti-acne medication may have been reponsible for approximately 1,000 birth defects in children born to women taking the drug during the period 1982–1986.[14] Exposure of the fetus to the drug over a period of only several days can result in birth defects such as severe facial malformations, heart defects, and mental retardation.

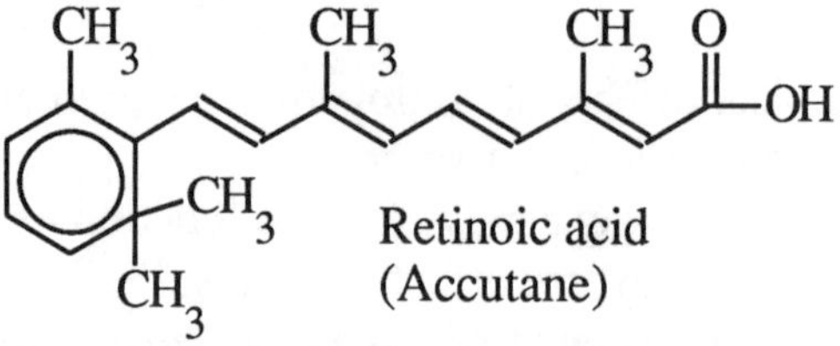

Retinoic acid
(Accutane)

4.8. XENOBIOTICS AND THE IMMUNE SYSTEM

The immune system response to toxicants, **immunogenesis**, was discussed in Section 3.10. Perhaps the most common biochemical response to a xenobiotic compound is the production of antibodies to xenobiotic compounds (antigens) bound as conjugates with proteins. In such a case the foreign body is called a hapten. Antibody production is the basis for the body's immunologic response to many small-molecule xenobiotics.

The immune response of animals to foreign substances is being put to increasing use as a means of determining the quantities of such substances in biological samples. The overall process involved is called **immunoassay**. There is not space here to discuss immunoassay in any detail. In general, it is based upon the production and isolation of antibodies specific for a particular chemical species (hapten) when that species is bound to a protein and introduced into the blood of a rabbit or other animal. The isolated and purified antibody produced can be used as an analytical reagent and has a very high specificity for the hapten when it is present in samples to be analyzed. When the appropriate separation and detection methods are used, very sensitive analyses can be performed based on this interaction.[15,16]

4.9. RADIATION

One of the greater concerns of modern civilization is that of the effects of ionizing radiation on living systems. The toxicologic effects of radiation have to do with its physical and chemical interactions with matter and the biological consequences that result.[17] Ionizing radiation alters chemical species in tissue and can lead to significant and harmful alterations in the tissue and in the cells that make up the tissue. Radon and radium, two radioactive elements of particular concern for their potential to expose humans to ionizing radiation, are discussed in Chapter 5.

LITERATURE CITED

1. Hodgson, Ernest, and Walter C. Dauterman, "Metabolism of Toxicants — Phase I Reactions," Chapter 4, and Walter C. Dauterman, "Metabolism of Toxicants," Chapter 5, in *Introduction to Biochemical Toxicology*, Ernest Hodgson and Frank E. Guthrie, Eds., Elsevier, New York, 1980, pp. 67–105.

2. Rawn, J. David, *Biochemistry*, Harper and Row Publishers, New York, 1983.

3. Timbrell, John A., *Principles of Biochemical Toxicology*, Taylor and Francis, Ltd., London, 1982.

4. Williams, Gary M., and John H. Weisburger, "Chemical Carcinogens," Chapter 5 in *Casarett and Doull's Toxicology*,

3rd ed., Curtis D. Klaassen, Mary O. Amdur, and John Doull, Eds., Macmillan Publishing Co., New York, 1986, pp. 99–173.

5. Sipes, Glenn and A. Jay Gandolfi, "Biotransformation of Toxicants," I. Chapter 4 in *Casarett and Doull's Toxicology*, 3rd ed., Curtis D. Klaassen, Mary O. Amdur, and John Doull, Eds., Macmillan Publishing Co., New York, 1986, pp. 64–98.

6. "Environmental Biochemistry and Chemical Toxicology," Chapter 18 in *Environmental Chemistry*, 4th ed., Stanley E. Manahan, Brooks/Cole Publishing Co., Monterey, CA, 1984.

7. Thilly, William G., and Katherine M. Call, "Genetic Toxicology," Chapter 6 in *Casarett and Doull's Toxicology*, 3rd ed., Curtis D. Klaassen, Mary O. Amdur, and John Doull, Eds., Macmillan Publishing Co., New York, 1986, pp. 174–194.

8. "Children Absorb Tris-BP Flame Retardant through Sleepware: Urine Contains the Mutagenic Metabolite, 2,3-Dibromopropanol," *Science*, **201**, 1020–23 (1978).

9. Diamond, Leila , "Tumor Promoters and Cell Transformation," Chapter 3 in *Mechanisms of Cellular Transformation by Carcinogenic Agents*, D. Grunberger and S. P. Goff, Eds., Pergamon Press, New York, 1987, pp. 73–132.

10. Holbrook, David J. , Jr., "Chemical Carcinogenesis," Chapter 16 in *Introduction to Biochemical Toxicology*, Ernest Hodgson and Frank E. Guthrie, Eds., Elsevier, New York, 1980, pp. 310–329.

11. Levi, Patricia E.. "Toxic Action," Chapter 6 in *Modern Toxicology*, Ernest Hodgson and Patricia E. Levi, Eds., Elsevier, New York, 1987, pp. 133–184.

12. "The Detection of Environmental Mutagens and Potential Carcinogens," Bruce N. Ames, *Cancer*, **53**, 2034–2040 (1984).

13. R. B. Kurzel and C. L. Cetrulo, "The Effect of Environmental Pollutants on Human Reproduction, Including Birth Defects," *Environmental Science and Technology*, **15**, 626-640 (1981).

14. "Anti-Acne Drug Faulted in Birth Defects," *New York Times*, April 22, 1988, p. 1.

15. Collins, W. P., *Alternative Immunoassays*, John Wiley and Sons, New York, 1985.

16. Van Emon, Jeanette, Bruce Hammock, and James N. Sieber, "Enzyme-Linked Immunosorbent Assay for Paraquat and its Application to Exposure Analysis," *Analytical Chemistry*, **58**, 1866–1873 (1986).

17. Paic, Guy, *Ionizing Radiation: Protection and Dosimetry*, CRC Press, Boca Raton, Florida, 1986.

5

Toxic Elements

5.1. INTRODUCTION

It is somewhat difficult to define what is meant by a toxic element. Some elements, such as white phosphorus, chlorine, and mercury, are quite toxic in the elemental state. Others, such as carbon, nitrogen and oxygen, are harmless as usually encountered in their normal elemental forms. With the exception of those noble gases that are not known to combine chemically, all elements can form toxic compounds. A prime example is hydrogen cyanide. This extremely toxic compound is formed from three elements that are nontoxic in the uncombined form, but which produce compounds that are essential constituents of living matter.

However, there are some elements that are notable for the toxicities of most of their compounds. A number of metals form very toxic ions. The elemental forms of some elements are extremely toxic. These three categories of elements are discussed in this chapter as toxic elements, with the qualification that this category is somewhat arbitrary. With the exceptions of phosphorus and chlorine, elements known to be essential to life processes in humans have not been included as toxic elements.

5.2. TOXIC ELEMENTS AND THE PERIODIC TABLE

It is most convenient to consider elements from the perspective of the periodic table, which is shown in Figure 2.1 and discussed in Section 2.3. Recall that the three main types of elements, based upon their chemical and physical properties as determined by the electron configurations of their atoms, are metals, nonmetals, and metalloids. Metalloids (B, Si, Ge, As, Sb, Te, At), show some characteristics of both metals and nonmetals. The nonmetals consist of those few elements in groups 4A–7A to the above and to the right of the metalloids. The noble gases, only some of which form a limited number of very unstable chemical compounds of no toxicological significance, are in group 8A. All the remaining elements, including the lanthanide and actinide series, are metals. Elements in the periodic table are broadly distinguished between representative elements in the A groups of the periodic table and transition metals constituting the B groups, the lanthanide series, and the actinide series.

5.3. ESSENTIAL ELEMENTS

Some elements are essential to the composition or function of the body. Since the body is mostly water, hydrogen and oxygen are obviously essential elements. Carbon (C) is a component of all life molecules, including proteins, lipids, and carbohydrates. Nitrogen, N, is in all proteins. The other essential nonmetals are phosphorus (P), sulfur (S), chlorine (Cl), selenium (Se), fluorine (F), and iodine (I). The latter two are among the essential trace elements that are required in only small quantities, particularly as constituents of enzymes or as cofactors (nonprotein species essential for enzyme function).[1] The metals present in macro amounts in the body are sodium (Na), potassium (K), and calcium (Ca). Essential trace elements are chromium (Cr), manganese (Mn), iron (Fe), cobalt (Co), copper (Cu), zinc (Zn), magnesium (Mg), molybdenum (Mo), nickel (Ni), and perhaps one or more elements that have not yet been established as essential.

5.4. METALS IN AN ORGANISM

Metals in the body are almost always in an oxidized or chemically combined form; mercury is a notable exception in that elemental mercury vapor readily enters the body through the pulmonary route. The simplest form of a chemically bound metal in the body is the hydrated cation, of which $Na(H_2O)_6^+$ is the most abundant example in the body. At pH values ranging upward from somewhat less than 7 (neutrality), many metal ions tend to be bound to one or more hydroxide groups; an example is iron(II) in $Fe(OH)(H_2O)_5^+$. Some metal ions have such a strong tendency to lose H^+ that, except at very low pH values, they exist as the insoluble hydroxides. A common example of this phenomenon is iron(III) which is very stable as the insoluble hydrated iron(III) oxide, $Fe_2O_3{\cdot}xH_2O$, or hydroxide, $Fe(OH)_3$. Metals can bond to some anions in body fluids. For example, in the strong hydrochloric acid medium of the stomach, some iron(III) may be present as $HFeCl_4$, where the acid in the stomach prevents formation of insoluble $Fe(OH)_3$ and a high concentration of chloride ion is available to bond to iron(III). Ion pairs may exist that consist of positively charged metal cations and negatively charged anions endogenous to body fluids. These do not involve covalent bonding between cations and anions, but rather an electrostatic attraction, such as in the ion pairs, $Ca^{2+}HCO_3^-$ or $Ca^{2+}Cl^-$.

Complex Ions and Chelates

With the exception of group 1A metals and the somewhat lesser exception of group 2A metals, there is a tendency for metals to form **complexes** with **electron donor** functional groups on **ligands** consisting of anionic or neutral inorganic or organic species. In such cases, covalent bonds are formed between the **central metal ion** and the ligands. Usually the resulting complex has a net charge and is called a complex ion; $FeCl_4^-$ is such an ion. In many cases an organic ligand has two or more electron donor functional groups that may simultaneously bond to a metal ion to form a complex with one or more rings in its structure. A ligand with this capability is called a **chelating agent**, and the complex is a **metal chelate.** Copper(II) ion forms such a chelate with the anion of the amino acid glycine as shown in Figure 5.1. This chelate is very stable.

Organometallic compounds constitute a large class of metal-containing species with properties quite different from those of the metal ions. These are compounds in which the metal is covalently bonded to carbon in an organic moiety, such as the methyl group, $-CH_3$. Unlike metal complexes, which can reversibly dissociate to the metal ions and ligands, the organic portions of organometallic compounds are not normally stable by themselves. The chemical and toxicological properties of organometallic compounds are discussed in detail in Chapter 6, so space will not be devoted to them here. However, it should be mentioned that neutral organometallic compounds tend to be lipid-soluble, a property that enables their facile movement across biologic membranes.[2] They often remain intact during movement through biological systems and so become distributed in these systems as lipid-soluble compounds.

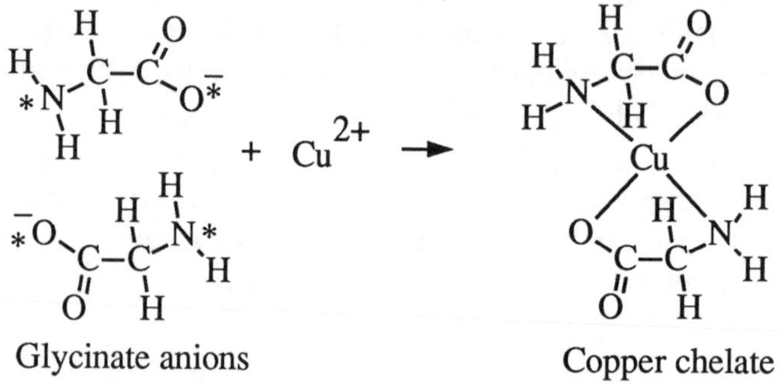

Glycinate anions + Cu^{2+} → Copper chelate

Figure 5.1. Chelation of Cu^{2+} by glycinate anion ligands to form the glycinate chelate. Each electron donor group on the glycinate anion chelating agents is designated with an asterisk. In the chelate the central copper(II) metal ion is bonded in 4 places and the chelate has two rings composed of the 5-atom sequence Cu-O-C-C-N.

A phenomenon not confined to metals, **methylation** is the attachment of a methyl group to an element and is a significant natural process responsible for much of the environmental mobility of some of the heavier elements.[3] Among the elements for which methylated forms are found in the environment are cobalt, mercury, silicon, phosphorus, sulfur, the halogens, germanium, arsenic, selenium, tin, antimony, and lead.

Metal Toxicity

Inorganic forms of most metals tend to be strongly bound by protein and other biologic tissue. Such binding increases bioaccumulation and inhibits excretion. There is a significant amount of tissue selectivity in the binding of metals. For example, toxic lead and radioactive radium are accumulated in osseous (bone) tissue, whereas the kidneys accumulate cadmium and mercury. Metal ions most commonly bond with amino acids, which may be contained in proteins (including enzymes) or polypeptides. The electron-donor groups most available for binding to metal ions are amino and carboxyl groups (see Figure 5.2). Binding is especially strong for many metals to thiol (sulfhydryl) groups, which is especially significant because the –SH groups are common components of the active sites of many crucial enzymes, including those that are involved in cellular energy output and oxygen transport. The amino acid that usually provides –SH groups in enzyme active sites is cysteine, as shown in Figure 5.2. The imidazole group of the amino acid histidine is a common feature of enzyme active sites with strong metal binding capabilities.[1]

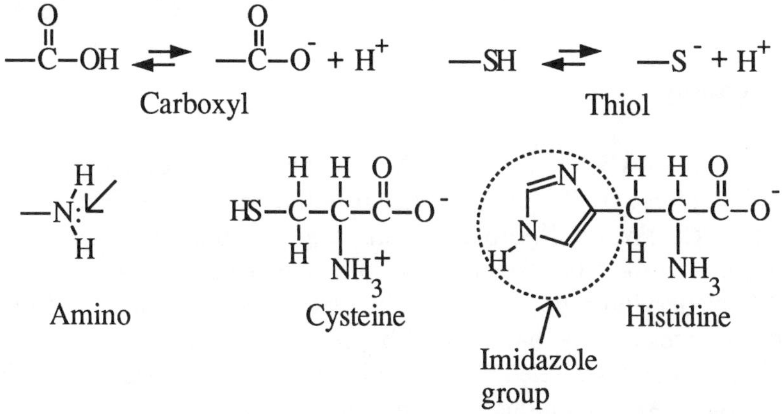

$$-C(=O)-OH \rightleftarrows -C(=O)-O^- + H^+ \qquad -SH \rightleftarrows -S^- + H^+$$

Carboxyl Thiol

$$-NH_2 \qquad HS-CH_2-CH(NH_3^+)-C(=O)-O^- \qquad \text{(imidazole)}-CH_2-CH(NH_3)-C(=O)-O^-$$

Amino Cysteine Histidine

Imidazole group

Figure 5.2. Major binding groups for metal ions in biologic tissue (carboxyl, thiol, amino) and amino acids with strong metal binding groups in enzyme active sites (cysteine, histidine). The arrow pointing to the amino group designates an unshared pair of electrons available for binding metal ions. The thiol group is a weak acid that usually remains unionized until the hydrogen ion is displaced by a metal ion.

The absorption of metals is to a large extent a function of their chemical form and properties. Pulmonary intake results in the most facile absorption and rapid distribution through the circulatory system. Absorption through this route is often very efficient when the metal is in the form of respirable particles of less than 100 μm in size, as volatile organometallic compounds (see Chapter 6) or (in the case of mercury) as the elemental metal vapor. Absorption through the gastrointestinal tract is affected by pH, rate of movement through the tract, and presence of other materials. Particular combinations of these factors can combine to make absorption very high or very low.

Metals tend to accumulate in target organs, and a toxic response is observed when the level of the metal in the organ reaches or exceeds a threshold level. Often the organs most affected are those involved with detoxication or elimination of the metal. Therefore, the liver and kidneys are often affected by metal poisoning. The form of the metal can determine which organ is adversely affected. For example, lipid-soluble elemental or organometallic mercury damages the brain and nervous system, whereas Hg^{2+} ion may attack the kidneys.

Because of the widespread opportunity for exposure combined with especially high toxicity, some metals are particularly noted for their toxic effects. These are discussed separately in the following sections in the general order of their appearance in groups in the periodic table.

Beryllium

Beryllium (Be) is a group 2A element with the electron configuration {He}$2s^2$. It is the first metal in the periodic table to be notably toxic. When fluorescent lamps and neon lights were first introduced, they contained beryllium phosphor, and a number of cases of beryllium poisoning resulted from the manufacture of these light sources and the handling of broken lamps. Modern uses of beryllium in ceramics, electronics, and alloys require special handling procedures to avoid industrial exposure.

Beryllium has a number of toxic effects. Of these, the most common involve the skin. Skin ulceration and granulomas have resulted from exposure to beryllium. Hypersensitization to beryllium can result in skin dermatitis, acute conjunctivitis, and corneal laceration.

Chronic berylliosis may occur with a long latent period of 5–20 years. The most damaging effect of chronic berylliosis is lung fibrosis and pneumonitis. In addition to coughing and chest pain, the subject suffers from fatigue, weakness, loss of weight, and dyspnea (difficult, painful breathing). The impaired lungs do not transfer oxygen well. Other organs that can be adversely affected are the liver, kidneys, heart, spleen, and striated muscles.

The chemistry of beryllium is atypical compared to that of the other group 1A and group 2A metals. Atoms of Be are the smallest of all metals, having an atomic radius of 111 pm (picometers). The beryllium ion, Be^{2+}, has an ionic radius of only 35 pm, which gives it a high polarizing ability, a tendency to form molecular compounds rather than ionic compounds, and a much greater tendency to form complex compounds than other group 1A or 2A ions.[4]

Vanadium

Vanadium (V) is a transition metal with an electron configuration of {Ar}$4s^2 3d^3$. In the combined form it exists in the +3, +4, and +5 oxidation states, of which the +5 is the most common. Vanadium is of concern as an environmental pollutant because of its high levels in residual fuel oils and subsequent emission as small particulate matter from the combustion of these oils in urban areas. Vanadium occurs as chelates of the porphyrin type in crude oil and it concentrates in the higher boiling fractions during the refining process. A major industrial use of vanadium is in catalysts, particularly those in which sulfur dioxide is oxidized in the production of sulfuric acid. The other major industrial uses of vanadium are for hardening steel, as a pigment ingredient, in photography, and as an ingredient of some insecticides. In addition to environmental exposure from the combustion of vanadium-containing fuels, there is some potential for industrial exposure.

Probably the vanadium compound to which people are most likely to be exposed is vanadium pentoxide, V_2O_5. Exposure normally occurs via the respiratory route, and the pulmonary system is the most likely to suffer from vanadium toxicity. Bronchitis and bronchial pneumonia are the most common pathological effects of exposure; skin and eye irritation may also occur. Severe exposure can also adversely affect the gastrointestinal tract, kidneys, and nervous system.

Chromium

Chromium (Cr) is a transition metal with an electron configuration of {Ar}$4s^1 3d^5$. In the chemically combined form it exists in all oxidation states from +2 through +6, of which the +3 and +6 are the more notable.

In strongly acidic aqueous solution, chromium(III) may be present as the hydrated cation $Cr(H_2O)_6^{3+}$. At pH values above approximately 4 this ion has a strong tendency to precipitate from solution as shown by the reaction:

$$Cr(H_2O)_6^{3+} \rightarrow Cr(OH)_3(s) + 3H^+ + 3H_2O \quad (5.1)$$

The two major forms of chromium(VI) in solution are the yellow chromate, CrO_4^{2-}, and orange dichromate, $Cr_2O_7^{2-}$. The latter predominates in acidic solution as shown by the following reaction, the equilibrium of which is forced to the left by higher levels of H^+:

$$Cr_2O_7^{2-} + H_2O \rightleftarrows 2HCrO_4^- \rightleftarrows 2H^+ + 2CrO_4^{2-} \quad (5.2)$$

Chromium in the +3 oxidation state is an essential trace element (see Section 5.3) required for glucose and lipid metabolism in mammals, and a deficiency of it gives symptoms of diabetes mellitus. However, chromium must also be discussed as a toxicant because of its toxicity in the +6 oxidation state, commonly called **chromate**. Exposure to chromium(VI) usually involves chromate salts, such as Na_2CrO_4. These salts tend to be water-soluble and readily absorbed into the bloodstream through the lungs. The carcinogenicity of chromate has been demonstrated by studies of exposed workers. Exposure to atmospheric chromate may cause bronchogenic carcinoma with a latent period of 10–15 years. In the body, chromium(VI) is readily reduced to chromium(III) as shown in Reaction 5.3; however, the reverse reaction does not occur in the body.

$$CrO_4^{2-} + 8H^+ + 3e^- \rightarrow Cr^{3+} + 4H_2O \quad (5.3)$$

Cadmium

Along with mercury and lead, cadmium (Cd) is one of the "big three" heavy metal poisons. Cadmium occurs as a constituent of lead and zinc ores, from which it can be extracted as a by-product. Cadmium is used to electroplate metals to prevent corrosion, as a pigment, as a constituent of alkali storage batteries, and in the manufacture of some plastics.

Cadmium is located at the end of the second row of transition elements and has the electron configuration $\{Kr\}5s^2 4d^{10}$. The two outer *s* electrons in cadmium are the only ones involved in bonding, and the +2 oxidation state of the element is the only one exhibited in its compounds. In its compounds, cadmium occurs as the Cd^{2+} ion.

Cadmium is directly below zinc in the periodic table and behaves much like zinc. This may account in part for cadmium's toxicity; because zinc is an essential trace element, cadmium substituting for zinc could cause metabolic processes to go wrong.

The toxic nature of cadmium was revealed in the early 1900s as a result of workers inhaling cadmium fumes or dusts in ore processing and manufacturing operations. Welding or cutting metals plated with cadmium or containing cadmium in alloys, or the use of cadmium rods or wires for brazing or silver-soldering, can be a particularly dangerous route to pulmonary exposure.[5] Acute pulmonary symptoms of cadmium exposure are usually caused by the inhalation of cadmium oxide dusts and fumes, which results in cadmium pneumonitis characterized by edema and pulmonary epithelium necrosis. Chronic exposure sometimes produces emphysema severe enough to be disabling. The kidney is generally regarded as the organ most sensitive to chronic cadmium poisoning. The function of renal tubules is impaired by cadmium as manifested by excretion of both high molecular mass proteins (such as albumin) and low molecular mass proteins.

The most spectacular and publicized occurrence of cadmium poisoning resulted from dietary intake of cadmium by people in the Jintsu River Valley near Fuchu, Japan. The victims were afflicted by *itai, itai* disease, which means "ouch, ouch" in Japanese. The symptoms are the result of painful osteomalacia (bone disease) combined with kidney malfunction. Cadmium poisoning in the Jintsu River Valley was attributed to irrigated rice contaminated from an upstream mine producing lead, zinc, and cadmium.

In general, cadmium is poorly absorbed through the gastrointestinal tract. A mechanism exists for its active absorption in the small intestine through the action of the low-molecular-mass calcium-binding protein CaBP. The production of this protein is stimulated by a calcium-deficient diet, which may aggravate cadmium toxicity.

Cadmium is a highly **cumulative** poison with a biologic half-life estimated at about 20-30 years in humans. About half of the body burden of cadmium is found in the liver and kidneys. The total body burden reaches a plateau in humans around age 50. Cigarette smoke is a source of cadmium, and the body burden of cadmium is about 1.5 to 2 times greater in smokers than in nonsmokers of the same age.

Cadmium in the body is known to affect several enzymes. It is believed that the renal damage that results in proteinuria from cadmium is the result of cadmium adversely affecting enzymes responsible for reabsorption of proteins in kidney tubules.[6] Cadmium also reduces the activity of delta-aminolevulinic acid synthetase (Figure 5.3), arylsulfatase, alcohol dehydrogenase, and lipoamide dehydrogenase, whereas it enhances the activity of delta-aminolevulinic acid dehydratase, pyruvate dehydrogenase, and pyruvate decarboxylase.

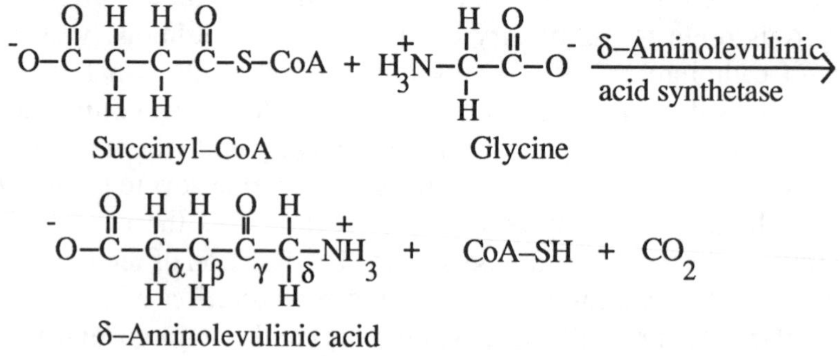

$$^{-}O{-}C(=O){-}CH_2{-}CH_2{-}C(=O){-}S{-}CoA \;(\text{Succinyl–CoA}) + H_3\overset{+}{N}{-}CH_2{-}C(=O){-}O^{-} \;(\text{Glycine}) \xrightarrow[\text{acid synthetase}]{\delta\text{–Aminolevulinic}}$$

$$^{-}O{-}C(=O){-}\underset{\alpha}{C}H_2{-}\underset{\beta}{C}H_2{-}\underset{\gamma}{C}(=O){-}\underset{\delta}{C}H_2{-}\overset{+}{N}H_3 \;(\delta\text{–Aminolevulinic acid}) + CoA{-}SH + CO_2$$

Figure 5.3. Path of synthesis of delta-aminolevulinic acid (coenzyme A abbreviated as CoA). Cadmium tends to inhibit the enzyme responsible for this process.

An interesting feature of cadmium metabolism is the role of **metallothionein**, which consists of two similar proteins with a low molecular mass of about 6,500. As a consequence of a high content of the amino acid cysteine:

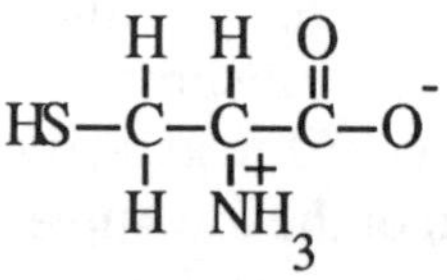

$$HS{-}CH_2{-}CH(\overset{+}{N}H_3){-}C(=O){-}O^{-} \quad \text{Cysteine}$$

metallothionein contains a large number of thiol (sulfhydryl, –SH) groups. These groups bind very strongly to other heavy metals, particularly mercury, silver, zinc, and tin. The general reaction of metallothionein with cadmium ion is the following:

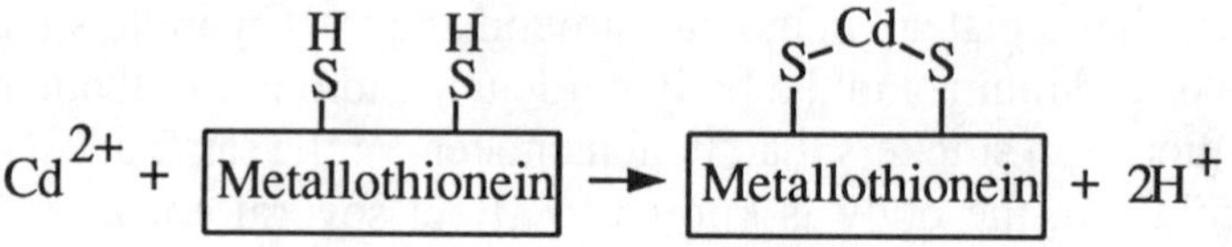

$$Cd^{2+} + \boxed{\text{Metallothionein}}(SH)_2 \rightarrow \boxed{\text{Metallothionein}}(S{-}Cd{-}S) + 2H^{+} \quad (5.4)$$

Metallothionein has been isolated from virtually all of the major mammal organs, including liver, kidney, brain, heart, intestine, lung, skin, and spleen. Nonlethal doses of cadmium, mercury, and lead induce synthesis of metallothionein. In test animals, nonlethal doses of cadmium followed by an increased level of metallothionein in the body have allowed later administration of doses of cadmium at a level fatal to nonacclimated animals, but without fatalities in the test subjects.

Cadmium is excreted from the body in both urine and feces. The mechanisms of cadmium excretion are not well known.

Mercury

Mercury is directly below cadmium in the periodic table, but has a considerably more varied and interesting chemistry than cadmium or zinc. Elemental mercury is the only metal that is a liquid at room temperature, and its relatively high vapor pressure contributes to its toxicological hazard. Mercury metal is used in electric discharge tubes (mercury lamps), gauges, pressure-sensing devices, vacuum pumps, valves, and seals. It was formerly widely used as a cathode in the chlor-alkali process for the manufacture of NaOH and Cl_2, a process that has been largely discontinued in part because of mercury pollution resulting from it.

In addition to the uses of mercury metal, mercury compounds have a number of applications.[7] Mercury(II) oxide, HgO, is commonly used as a raw material for the manufacture of other mercury compounds. Mixed with graphite, it is a constituent of the Ruben-Mallory dry cell for which the cell reaction is the following:

$$Zn + HgO \rightarrow ZnO + Hg \quad (5.5)$$

Mercury(II) acetate, $Hg(C_2H_3O_2)_2$ is made by dissolving HgO in warm 20% acetic acid. This compound is soluble in a number of organic solvents. Mercury(II) chloride is quite toxic. The dangers of exposure to $HgCl_2$ are aggravated by its high water solubility and relatively high vapor pressure compared to other salts. Mercury(II) fulminate, $Hg(ONC)_2$, is used as an explosives' detonator. In addition to the +2 oxidation state, mercury can also exist in the +1 oxidation state as the dinuclear Hg_2^{2+} ion. The best known mercury(I) compound is mercury(I) chloride, Hg_2Cl_2, commonly called calomel. It is a constituent of calomel reference electrodes, such as the well known saturated calomel electrode, SCE.

A number of organomercury compounds are known. These compounds and their toxicities are discussed further in Chapter 6.

Absorption and Transport of Elemental and Inorganic Mercury

Monatomic elemental mercury in the vapor state, Hg(*g*), is absorbed from inhaled air by the pulmonary route to the extent of about 80%. Inorganic mercury compounds are absorbed through the intestinal tract and in solution through the skin.

Although elemental mercury is rapidly oxidized to mercury(II) in erythrocytes (red blood cells), a large fraction of elemental mercury absorbed through the pulmonary route reaches the brain prior to oxidation and enters that organ because of the lipid solubility of mercury(0). This mercury is subsequently oxidized in the brain and remains there. Inorganic mercury(II) tends to accumulate in the kidney.

Metabolism, Biologic Effects, and Excretion

Like cadmium, mercury(II) has a strong affinity for sulfhydryl groups in proteins, enzymes, hemoglobin, and serum albumin. Because of the abundance of sulfhydryl groups in active sites of many enzymes, it is difficult to establish exactly which enzymes are affected by mercury in biological systems.

The effect upon the central nervous system following inhalation of elemental mercury is largely psychopathological. Among the most prominent symptoms are tremor (particularly of the hands) and emotional instability characterized by shyness, insomnia, depression, and irritability. These symptoms are probably the result of damage to the blood-brain barrier. This barrier regulates the transfer of metabolites, such as amino acids, to and from the brain. Brain metabolic processes are probably disrupted by the effects of mercury.

The kidney is the primary target organ for Hg^{2+}. Chronic exposure to inorganic mercury(II) compounds causes proteinuria. In cases of mercury poisoning of any type, the kidney is the organ with the highest bioaccumulation of mercury.

Excretion of inorganic mercury occurs through the urine and

feces. The mechanisms by which excretion occurs are not well understood.

Minimata Bay

The most notorious incident of widespread mercury poisoning in modern times occurred in the Minimata Bay region of Japan during the period 1953-1960. Mercury waste from a chemical plant draining into the bay contaminated seafood consumed regularly by people in the area. Overall, 111 cases of poisoning with 43 deaths and 19 congenital birth defects were documented. The seafood was found to contain 5-20 parts per million of mercury.

Lead

Lead (Pb) ranks fifth behind iron, copper, aluminum, and zinc in industrial production of metals.[8] About half of the lead used in the United States goes for the manufacture of lead storage batteries. Other uses include solders, bearings, cable covers, ammunition, plumbing, pigments, and caulking.

Metals commonly alloyed with lead are antimony (in storage batteries), calcium and tin (in maintenance-free storage batteries), silver (for solder and anodes), strontium and tin (as anodes in electrowinning processes), tellurium (pipe and sheet in chemical installations and nuclear shielding), tin (solders), and antimony and tin (sleeve bearings, printing, high-detail castings).

The electron configuration of lead is $\{Xe\}4f^{14}5d^{10}6s^{2}6p^{2}$ and it has four valence electrons. Lead is one of a small group of "p-block" metals that are representative elements with partially filled *p* orbitals. Inorganic lead exists in chemical compounds in both the +2 and +4 oxidation states. Lead(II) compounds are predominantly ionic (e.g., $Pb^{2+}SO_4^{2-}$), whereas lead(IV) compounds tend to be covalent (e.g., tetraethyllead, $Pb(C_2H_5)_4$). Some lead(IV) compounds, such as PbO_2, are strong oxidants. Lead forms several basic lead salts, such as $Pb(OH)_2{\cdot}2PbCO_3$, which was once the most widely used white paint pigment and the source of considerable chronic lead poisoning to the children who ate peeling white paint. Many compounds of lead in the +2 oxidation state (lead(II)) and a few in the +4 oxidation state (lead(IV)) are useful. The two most common of these are lead dioxide and lead sulfate, which are participants in the following reversible reaction that occurs during the charge and discharge of a lead storage battery:

$$Pb + PbO_2 + 2H_2SO_4 \rightleftharpoons 2PbSO_4 + 2H_2O \qquad (5.6)$$

Charge ← → Discharge

Lead halides have several important uses, such as that of $PbCl_2$ in asbestos brake linings and clutch disks. Lead hydroxide, $Pb(OH)_2$, is a component of sealed nickel-cadmium batteries. Basic lead sulfates ($xPbO{\cdot}PbSO_4$, x = 1–4) are used as paint pigments. Dibasic lead phosphite ($2PbO_2{\cdot}PbHPO_3{\cdot}1/2H_2O$) is used as a stabilizer in polyvinylchloride plastic to give this polymer desired qualities of thermal stability, weathering resistance, and electrical insulation.

In addition to the inorganic compounds of lead, there are a number of organolead compounds. The manufacture of tetraethyllead used to rank second to storage batteries for lead use, but is now dropping because of severe restrictions on the uses of this antiknock additive in gasoline. Organolead compounds are discussed in Chapter 6.

Exposure and Absorption of Inorganic Lead Compounds

Although industrial lead poisoning used to be very common, it is relatively rare now[9] because of previous experience with the toxic effects of lead and protective actions that have been taken. From the preceding discussion, it is obvious that lead is widely used and distributed, so that opportunities for exposure of the general population are relatively abundant. Of all the toxic elements, lead is the one that the average person is most likely to encounter at a level detrimental to health.

Lead is a common atmospheric pollutant, and absorption through the respiratory tract is the most common route of human exposure. The greatest danger of pulmonary exposure comes from inhalation of very small respirable particles of lead oxide (particularly from lead smelters and storage battery manufacturing) and lead carbonates, halides, phosphates, and sulfates. The other major route of lead absorption is the gastrointestinal tract. Lead, which in the +2 oxidation state has some chemical similarities to calcium, may have the same transport mechanism as this element in the gastrointestinal tract. It is known that lead absorption decreases with increased levels of calcium in the diet and vice versa.

Transport and Metabolism of Lead

A striking aspect of lead in the body is its very rapid transport to bone and storage there. Lead tends to undergo bioaccumulation in bone throughout life, and about 90% of the body burden of lead is in bone after long-term exposure. Of the soft tissues, the liver and kidney tend to have somewhat elevated lead levels.

Measurement of the concentration of lead in the blood is the standard test for recent or ongoing exposure to lead. This test is used routinely to monitor industrial exposure to lead and in screening children for lead exposure.

The most common biochemical effect of lead is inhibition of the synthesis of heme, a complex (see Section 6.4) of a substituted porphyrin and Fe^{2+} in hemoglobin and cytochromes. Lead interferes with the conversion of delta-aminolevulinic acid to porphobilinogen, as shown in Figure 5.4, with a resulting accumulation of metabolic products. Hematological damage results. Lead inhibits enzymes that have sulfhydryl groups. However, the affinity of lead for the –SH group is not as great as that of cadmium or mercury.

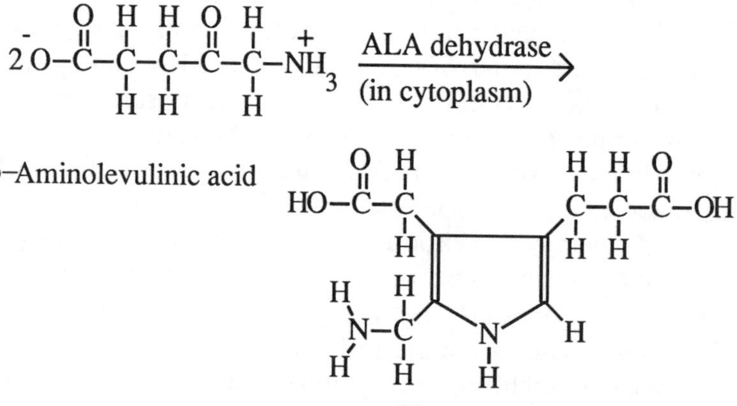

$$2\ {}^{-}O-C(=O)-CH_2-CH_2-C(=O)-CH_2-NH_3^{+} \xrightarrow[\text{(in cytoplasm)}]{\text{ALA dehydrase}}$$

δ–Aminolevulinic acid

Porphobilinogen

Figure 5.4. Synthesis of porphobilinogen from δ-aminolevulinic acid, a major step in the overall scheme of heme synthesis that is inhibited by lead in the body.

Manifestations of Lead Poisoning

Lead adversely affects a number of systems in the body. The inhibition of the synthesis of hemoglobin by lead has just been noted. This effect, plus a shortening of the life span of erythrocytes, results in anemia, a major manifestation of lead poisoning.

The central nervous system is adversely affected by lead. Psychopathological symptoms include restlessness, dullness, irritability, and memory loss. The subject may experience ataxia, headaches, and muscular tremor. In extreme cases, convulsions followed by coma and death may occur. Lead affects the peripheral nervous system, and lead palsey used to be a commonly observed symptom in lead industry workers and miners suffering from lead poisoning. Even in the absence of lead palsey, peripheral nerves are adversely affected by chronic lead poisoning.

Lead causes reversible damage to the kidney through its adverse effect upon proximal tubules. This impairs the processes by which the kidney absorbs glucose, phosphates, and amino acids prior to secretion of urine. A longer-term effect of lead ingestion on the kidney is general degradation of the organ (chronic nephritis) including glomular atrophy, interstitial fibrosis, and sclerosis of vessels.

Reversal of Lead Poisoning and Therapy

Some effects of lead poisoning, such as those upon proximal tubules of the kidney and inhibition of heme synthesis, are reversible upon removal of the source of lead exposure. Lead poisoning can be treated by chelation therapy in which the lead is solubilized and removed by a chelating agent (see Section 5.4). One such chelating agent is ethylenediaminetetraacetic acid (EDTA) which binds strongly to most +2 and +3 cations (Figure 5.5). It is administered for lead poisoning therapy in the form of the calcium chelate. The ionized Y^{4-} form chelates metal ions by bonding at one, two, three, or all four carboxylate groups ($-CO_3^{2-}$) and one or both of the 2 N atoms (see glycinate-chelated structure in Figure 5.1). EDTA is administered as the calcium chelate for the treatment of lead poisoning to avoid any net loss of calcium by solubilization and excretion.

Another compound used to treat lead poisoning is British anti-Lewisite (BAL), originally developed to treat arsenic-containing poison gas Lewisite. As shown in Figure 5.6, BAL chelates lead through its sulfhydryl groups, and the chelate is excreted through the kidney and bile.

```
        O H                          H O
   *    ‖ |                          | ‖   *
   HO—C—C                            C—C—OH
          |\                        /|
          H \    H     H           / H
             \   |     |          /
              *N—C—————C—N *
             /   |     |          \
          H /    H     H           \ H
   *      |/                        \|     *
   HO—C—C                            C—C—OH
      ‖ |                            | ‖
      O H                            H O
```

EDTA

Figure 5.5. The non-ionized form of ethylenediaminetetraacetic acid, EDTA. (Asterices denote binding sites.)

```
        H     H  H       (2-)
        |     |  |
      H—C—————C—C—OH
        |     |  |
        S     S  H
         \   /
          Pb
         /   \
        S     S  H
        |     |  |
      H—C—————C—C—OH
        |     |  |
        H     H  H
```

Figure 5.6. Lead chelated by the lead antidote British anti-Lewisite, BAL.

5.5. METALLOIDS: ARSENIC

Sources and Uses

Arsenopyrite and loellingite are both arsenic minerals that can be smelted to produce elemental arsenic. Both elemental arsenic and arsenic trioxide (As_2O_3) are produced commercially; the latter is the raw material for the production of numerous arsenic compounds. Elemental arsenic is used to make alloys with lead and copper.[10] Arsenic compounds have a number of uses, including applications in catalysts, bactericides, herbicides, fungicides, animal feed additives, corrosion inhibitors, pharmaceuticals, veterinary medicines, tanning agents, and wood preservatives. Arsenicals were the first drugs to be effective against syphilis and they are still used to treat amebic dysentery. Arsobal, or Mel B, an organoarsenical, is the most effective drug for the treatment of the neurological stage of African trypanosomiasis for which the infectious agents are *Trypanosoma gambiense* or *T. rhodesiense*.

Exposure and Absorption of Arsenic

Arsenic can be absorbed through both the gastrointestinal and pulmonary routes. Although the major concern with arsenic is its effect as a systemic poison, arsenic trichloride ($AsCl_3$) and the organic arsenic compound, Lewisite (used as a poison gas in World War I) can penetrate skin; both of these compounds are very damaging at the point of exposure and are strong vesicants (causes of blisters).[11] The common arsenic compound As_2O_3 is absorbed through the lungs and intestines. The degree of coarseness of the solid is a major factor in how well it is absorbed. Coarse particles of this compound tend to pass through the gastrointestinal tract and to be eliminated with the feces.

Arsenic occurs in the +3 and +5 oxidation states, and inorganic compounds in the +3 oxidation state are generally more toxic. The conversion to arsenic(V) is normally favored in the environment, which somewhat reduces the overall hazard of this element.

Arsenic is a natural constituent of most soils. It is found in a number of foods, particularly shellfish. The average adult ingests somewhat less than 1 milligram of arsenic per day through natural sources.

Metabolism, Transport, and Toxic Effects of Arsenic

Biochemically, arsenic acts to coagulate proteins, forms complexes with coenzymes, and inhibits the production of ATP. Like cadmium and mercury, arsenic is a sulfur-seeking element. Arsenic has some chemical similarities to phosphorus,[12] and it substitutes for phosphorus in some biochemical processes, with adverse metabolic effects. Figure 5.7 summarizes one such effect. The top reaction in the figure illustrates the enzyme-catalyzed synthesis of 1,3-diphosphoglycerate from glyceraldehyde 3-phosphate. The product undergoes additional reactions to produce adenosine triphosphate (ATP), an essential energy-yielding substance in body metabolism. When arsenite AsO_3^{3-} is present, it bonds to glyceraldehyde 3-phosphate to yield a product that undergoes nonenzymatic spontaneous hydrolysis. This prevents ATP formation.

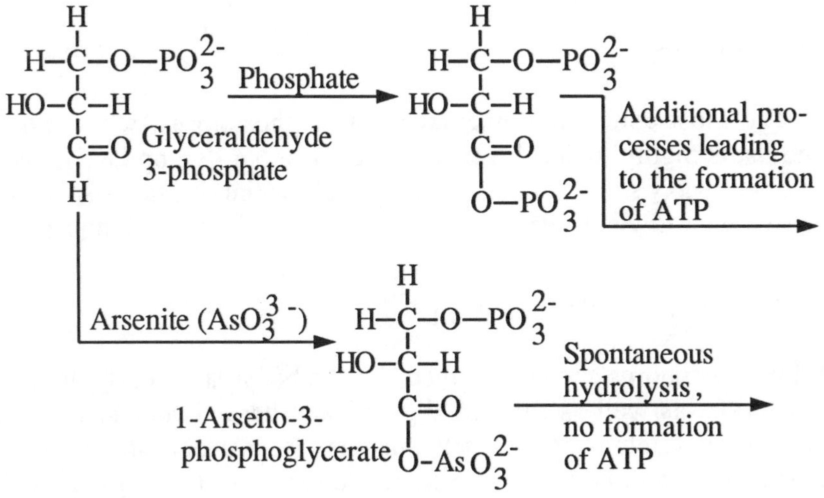

Glyceraldehyde 3-phosphate — Phosphate → Additional processes leading to the formation of ATP

Arsenite (AsO_3^{3-}) → 1-Arseno-3-phosphoglycerate → Spontaneous hydrolysis, no formation of ATP

Figure 5.7. Interference of arsenic(III) with ATP production by phosphorylation.

Antidotes to arsenic poisoning take advantage of the element's sulfur-seeking tendencies and contain sulfhydryl groups. One such antidote is 2,3-mercaptopropanol (BAL) discussed in the preceding section as an antidote for lead poisoning.

5.6. NONMETALS

Although molecular oxygen is essential for respiratory processes in aerobic organisms, it is reduced in the body to active species that can be harmful.[13] Successive additions of an electron (e^-) and a hydrogen ion (H^+) to a molecule of O_2 produce $HO_2\cdot$ (hydroperoxyl radical or superoxide), H_2O_2 (hydrogen peroxide), and $HO\cdot$ (hydroxyl radical, produced along with a molecule of H_2O). The dot beside the formulas $HO\cdot$ and $HO_2\cdot$ denotes that each one of these species contains an unpaired electron. Such species are called **radicals** and are very chemically reactive. These radicals and chemically reactive hydrogen peroxide attack tissue and DNA either directly or through their reaction products. The damage done is sometimes referred to as oxidative lesions. Radicals are scavenged from a living system by several enzymes, including peroxidase, superoxide dismutase, and catalase. Oxidative lesions on DNA may be repaired by DNA repair enzymes.

Phosphorous

The most common elemental form of phosphorus, white phosphorus, is highly toxic. White phosphorus (mp 44°C, bp 280°C) is a colorless waxy solid, sometimes with a yellow tint. It ignites spontaneously in air to yield a dense fog of finely divided, highly deliquescent P4O10:

$$P_4 + 5O_2 \rightarrow P_4O_{10} \tag{5.7}$$

White phosphorus can be absorbed into the body particularly through inhalation, as well as through the oral and dermal routes. It has a number of systemic effects, including anemia, gastrointestinal system dysfunction, and bone brittleness. Acute exposure to relatively high levels results in gastrointestinal disturbances and weakness due to biochemical effects on the liver. Chronic poisoning occurs largely through the inhalation of low concentrations of white phosphorus and causes necrosis (tissue death) of the jawbone, brittleness in other bones, and deterioration of teeth, which can result in their loss. **Phossy jaw** results from necrosis and fracture of the jawbone in exposed individuals. Severe eye damage can result from chronic exposure to elemental white phosphorus.

The Halogens

The elemental halogens — fluorine, chlorine, bromine, and iodine — are all toxic. Both fluorine and chlorine are highly corrosive gases that are very damaging to exposed tissue. All the halogens are used in their elemental forms and can properly be discussed with toxic elements. However, the toxicological properties of these elements are very similar in many cases to those of interhalogen compounds formed between the various halogens, which are covered in Chapter 7, and discussion of halogen toxicity is deferred to that chapter.

Radon and Radium

Radon

In section 4.9, the toxicological effects of ionizing radiation were mentioned, and radon was cited as a source of such radiation. Some authorities believe that of all the elemental toxicants radon is the one that is most likely to eventually cause death in humans.[14] The threat of radon in the environment and in indoor air have been summarized in books on the subject.[15,16] Radon's toxicity is not the result of its chemical properties, because it is a noble gas and does not enter into any normal chemical reactions. However, it is a radioactive element (radionuclide) that emits positively charged alpha particles, the largest and — when emitted inside the body — the most damaging form of radioactivity. Furthermore, the products of the radioactive decay of radon are also alpha emitters. Alpha particles emitted from a radionuclide in the lung cause damage to cells lining the lung bronchi and other tissues, resulting in processes that can cause cancer.

Radon is a decay product of radium which in turn is produced by the radioactive decay of uranium. During its brief lifetime, radon may diffuse upward through soil and into dwellings through cracks in basement floors. Radioactive decay products of radon become attached to particles in indoor air, are inhaled, and lodge in the lungs until they undergo radioactive decay, damaging lung tissue. Synergistic effects between radon and smoking appear to be responsible for most of the cases of cancer associated with radon exposure.

Radium

A second radionuclide to which humans are likely to be exposed is **radium**, Ra. Occupational exposure to radium is known to have caused cancers in humans. The most likely route for human exposures to low doses is through drinking water.[17] Areas in the United States where significant radium contamination of water has been observed include the uranium-producing regions of the western U.S., Iowa, Illinois, Wisconsin, Missouri, Minnesota, Florida, North Carolina, Virginia, and the New England states.

The maximum contaminant level (MCL) for total radium (^{226}Ra plus ^{228}Ra) in drinking water is specified by the U.S. Environmental Protection Agency as 5 pCi/L (picoCuries per liter) where a picoCurie is 0.037 disintegrations per second. Perhaps as many as several hundred municipal water supplies in the U.S. exceed this level and require additional treatment to remove radium. Fortunately, conventional water softening processes, which are designed to take out excessive levels of calcium, are relatively efficient in removing radium from water.

LITERATURE CITED

1. Hayes, J. A., "Metal Toxicity," Chapter 13 in *A Guide to General Toxicology*, Freddy Homburger, John A. Hayes and Edward W. Pelikan, Eds., Karger, New York, 1983, pp. 227-237.

2. Hammond, Paul B., and Robert P. Beliles, "Metals," Chapter 17 in *Casarett and Doull's Toxicology*, 2nd ed., John Doull, Curtis D. Klaassen, and Mary O. Amdur, Eds., Macmillan Publishing Co., New York, 1980, 409-467.

3. Thayer, John S., "Methylation: Its Role in the Environmental Mobility of Heavy Elements," *Preprint Extended Abstract*, American Chemical Society Division of Environmental Chemistry, **28**(1), American Chemical Society (1988), pp. 295-297.

4. Bailar, John C., Jr., Therald Moeller, Jacob Kleinberg, Cyrus O. Guss, Mary E. Castellion, and Clyde Metz, *Chemistry*, 2nd ed., Academic Press, Orlando, Florida, 1984.

5. Gosselin, Robert E., Roger P. Smith, and Harold C. Hodge, "Cadmium," in *Clinical Toxicology of Commercial Products* 5th ed., Williams and Wilkins, Baltimore, 1984, pp. III-77-III-84.

6. Donaldson, William E., "Trace Element Toxicity," Chapter 17 in *Introduction to Biochemical Toxicology*, Ernest Hodgson

and Frank E. Guthrie, Eds., Elsevier, New York, 1980, pp. 330-340.

7. Drake, Harold J., "Mercury," and William Singer and Milton Nowak, "Mercury Compounds" in *Kirk-Othmer Concise Encyclopedia of Chemical Technology*, Wiley-Interscience, New York, 1985, 744-745.

8. Howe, H., "Lead," and Dodd S. Carr, "Lead Compounds" in *Kirk-Othmer Concise Encyclopedia of Chemical Technology*, Wiley-Interscience, New York, 1985, pp. 688-692.

9. Ter Haar, Gary, "Industrial Toxicology (of lead)," in *Kirk-Othmer Concise Encyclopedia of Chemical Technology*, Wiley-Interscience, New York, 1985, p. 693.

10. Carapella, S. C., Jr., "Arsenic and Arsenic Alloys," in *Kirk-Othmer Concise Encyclopedia of Chemical Technology*, Wiley-Interscience, New York, 1985, pp. 134-135.

11. Gosselin, Robert E., Roger P. Smith, and Harold C. Hodge, "Arsenic," in *Clinical Toxicology of Commercial Products* 5th ed., Williams and Wilkins, Baltimore, 1984, pp. III-42–III-47.

12. Brown, Theodore L., and H. Eugene LeMay, Jr., *Chemistry — The Central Science*, Prentice-Hall, Englewood Cliffs, NJ, 1988.

13. Imlay, James A., Sherman M. Chin, and Stuart Linn, "Toxic DNA Damage by Hydrogen Peroxide through the Fenton Reaction *in Vivo* and *in Vitro*," *Science* **240,** 640-642 (1988).

14. Kerr, Richard A., "Indoor Radon: The Deadliest Pollutant," *Science*, **240**, 606-608 (1988).

15. Cothern, C. Richard, and James E. Smith, Jr., Eds., *Environmental Radon*, Plenum Press, New York, 1987.

16. Nazaroff, William W., and Anthony V. Nero, Jr., *Radon and its*

Decay Products in Indoor Air, John Wiley and Sons, New York, 1988.

17. Valentine, Richard L., Roger C. Splinter, Timothy S. Mulholland, Jeffrey M. Baker, Thomas M Nogaj and Jao-Jia Horng, *A Study of Possible Economical Ways of Removing Radium from Drinking Water*, EPA/600/S2-88/009, U. S. Environmental Protection Agency, Washington, DC, 1988.

6

Organometallics and Organometalloids

6.1. THE NATURE OF ORGANOMETALLIC AND ORGANOMETALLOID COMPOUNDS

An **organometallic compound** is one in which the metal atom is bonded to at least one carbon atom in an organic group. An **organometalloid compound** is a compound in which a metalloid element is bonded to at least one carbon atom in an organic group. The metalloid elements are shown in the periodic table of the elements in Figure 2.1 and consist of boron, silicon, germanium, arsenic, antimony, tellurium, and astatine (a very rare radioactive element). In subsequent discussion, *organometallic* will be used as a term to designate both organometallic and organometalloid compounds and *metal* will refer to both metals and metalloids, unless otherwise indicated. Given the predominance of the metals among the elements, and the ability of most to form organometallic compounds, it is not surprising that there are so many organometallic compounds, and new ones are being synthesized regularly. Fortunately, only a small fraction of these compounds are produced in nature or for commercial use, which greatly simplifies the study of their toxicities.

A further clarification of the nature of organometallic compounds is based upon the **electronegativities** of the elements involved, i.e., the abilities of covalently bonded atoms to attract electrons to themselves. Electronegativity values range from 0.86 for cesium to 4.10 for fluorine. The value for carbon is 2.50, and all organometallic compounds involve bonds between carbon and an element with an electronegativity value of less than 2.50. The value of the electronegativity of phosphorus is 2.06, but it is so nonmetallic in its behavior that its organic compounds are not classified as organometallic compounds.

6.2. CLASSIFICATION OF ORGANOMETALLIC COMPOUNDS

The simplest way to classify organometallic compounds for the purpose of discussing their toxicology is the following:[1]

1. Those in which the organic group is an alkyl group such as ethyl:

```
  H   H
  |   |
—C———C—H
  |   |
  H   H
```

   in tetraethyllead, $Pb(C_2H_5)_4$.

2. Those in which the organic group is carbon monoxide:

   :C≡O:

   (In the preceding Lewis formula of CO each dash, –, represents a pair of bonding electrons, and each pair of dots, :, represents an unshared pair of electrons.) Compounds with carbon monoxide bonded to metals, some of which are quite volatile and toxic, are called **carbonyls.**

3. Those in which the organic group is a π electron donor, such as ethylene or benzene.

```
H     H
 \   /
  C=C
 /   \
H     H
```

   Ethylene Benzene

Combinations exist of the three general types of compounds outlined above, the most prominent of which are arene carbonyl species in which a metal atom is bonded to both an aromatic entity such as benzene and to several carbon monoxide molecules. A more detailed discussion of the types of compounds and bonding follows.

Ionically Bonded Organic Groups

Negatively charged hydrocarbon groups are called **carbanions.** These can be bonded to group IA and IIA metal cations, such as Na^+ and Mg^{2+}, by predominantly ionic bonds. In some carbanions the negative charge is localized on a single carbon atom. For species in which conjugated double bonds and aromaticity are possible, the charge may be delocalized over several atoms, thereby increasing the carbanions's stability (see Figure 6.1).

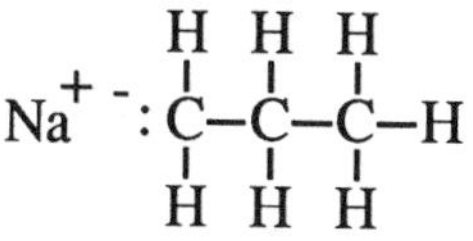

H H H
$Na^+ \; ^-$: C–C–C–H
H H H

Negative charge localized on a single carbon atom in propylsodium

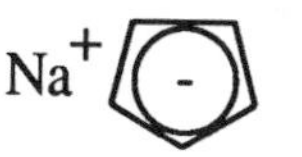

Na^+ (cyclopentadienide ring, -)

Negative charge delocalized in the 5-carbon ring of cyclopentadiene (see cyclopentadiene below)

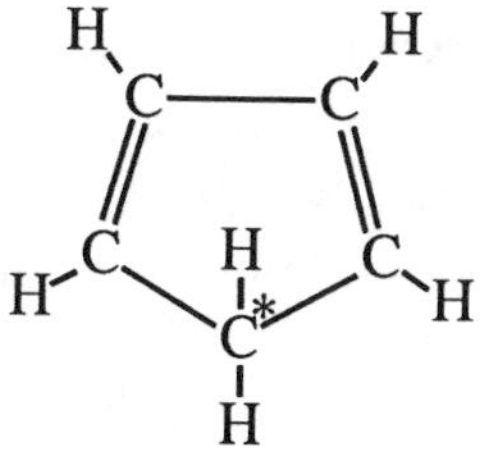

Cyclopentadiene. Loss of H^+ from the carbon marked with an asterisk gives the negatively charged cyclopentadienide anion.

Figure 6.1. Carbanions showing localized and delocalized negative charges.

Ionic organic compounds involving carbanions react readily with oxygen. For example, ethylsodium, $C_2H_5^-Na^+$, self-ignites in air. Ionic organometallic compounds are extremely reactive in water, as shown by the following reaction:

$$C_2H_5^-Na^+ \xrightarrow{H_2O} \text{Organic products} + NaOH \qquad (6.1)$$

One of the products of such a reaction is a strong base, such as NaOH, which is very corrosive to exposed tissue.

Organic Groups Bonded with Classical Covalent Bonds

A major group of organometallic compounds has carbon-metal

covalent single bonds in which both the C and metal (or metalloid) atoms contribute one electron each to be shared in the bond (in contrast to ionic bonds in which electrons are transferred between atoms). The bonds produced by this sharing arrangement are sigma-covalent bonds in which the electron density is concentrated between the two nuclei. Since in all cases the carbon atom is the more electronegative atom in this bond (see Section 6. 1), the electrons in the bond tend to be more attracted to the more electronegative atom and the covalent bond has a **polar** character as denoted by the following:

$$\overset{\delta+}{M}\text{——}\overset{\delta-}{C}$$

When the electronegativity difference is extreme, such as when the metal atom is Na, K, or Ca, an ionic bond is formed. In cases of less extreme differences in electronegativity, the bond may be only partially ionic; i.e., it is intermediate between a covalent and ionic bond. Organometallic compounds with classical covalent bonds are formed with representative elements and with zinc, cadmium, and mercury, which have filled *d* orbitals. In some cases these bonds are also formed with transition metals. Organometallic compounds with this kind of bonding comprise some of the most important and toxicologically significant organometallic compounds. Examples of such compounds are shown in Figure 6.2.

The two most common reactions of sigma-covalently bonded organometallic compounds are oxidation and hydrolysis (see Chapter 2). These compounds have very high heats of combustion because of the stabilities of their oxidation products, which consist of the metal oxide, water, and carbon dioxide as shown by the following reaction for the oxidation of dimethyl zinc:

$$Zn(C_2H_5)_2 + 7O_2 \rightarrow ZnO(s) + 5H_2O(g) + 4CO_2(g) \quad (6.2)$$

Industrial accidents in which the combustion of organometallic compounds generates respirable, toxic metal oxide fumes can certainly pose a hazard.

The organometallic compounds most likely to undergo hydrolysis are those with ionic bonds, compounds with relatively polar covalent bonds and those with vacant atomic orbitals (see Section 2.3) on the metal atom, which can accept more electrons. These provide sites of attack for the water molecules. For example, liquid trimethylaluminum reacts almost explosively with water or water and air:

$$Al(CH_3)_3 \xrightarrow[\{O_2\}]{H_2O} Al(OH)_3 + \text{Organic products} \qquad (6.3)$$

In addition to the dangers posed by the vigor of the reaction, it is possible that noxious organic products are evolved. Accidental exposure to air in the presence of moisture can result in the generation of sufficient heat to cause complete combustion of trimethylaluminum to the oxides of aluminum and carbon and to water.

```
        H
        |
      H-C-H
        |
        B
   H  /   \  H
    C       C
  H  H    H   H

  Trimethylboron


              H
              |
            H-C-H
      H       |       H
      |       |       |
    H-C------Si-------C-H
      |       |       |
      H       |       H
            H-C-H
              |
              H

        Tetramethylsilicon


          H
          |
        H-C-H
          |
        H-C-H
  H H     |     H H       H H H     H H H
  | |     |     | |       | | |     | | |
H-C-C----Pb-----C-C-H   H-C-C-C-Zn-C-C-C-H
  | |     |     | |       | | |     | | |
  H H     |     H H       H H H     H H H
        H-C-H
          |               Di-n-propylzinc
        H-C-H
          |
          H   Tetraethyllead
```

Figure 6.2. Some organometallic compounds with sigma-covalent metal-carbon bonds.

Organometallic Compounds with Dative Covalent Bonds

Dative covalent bonds, or coordinate covalent bonds, are those in which electrons are shared (as in all covalent bonds) but in which both electrons involved in each bond are contributed from the same atom. Such bonds occur in organometallic compounds of transition metals having vacant *d* orbitals. It is beyond the scope of this book to discuss such bonding in detail and the reader needing additional information is referred to works on organometallic compounds.[1,2] The most common organometallic compounds that have dative covalent bonds are **carbonyl compounds**, which are formed from a transition metal and carbon monoxide, where the metal is usually in the -1, 0, or +1 oxidation state. In these compounds the carbon atom on the carbon monoxide acts as an electron-pair donor as shown by the following:

$$M \; + \; :CO: \longrightarrow \underset{\uparrow}{M:CO:} \qquad (6.4)$$

Dative bond

Most carbonyl compounds have several carbon monoxide molecules bonded to a metal.

Many transition metal carbonyl compounds are known. The one of these that is the most significant toxicologically because of its widespread occurrence and extremely poisonous nature is the nickel carbonyl compound, $Ni(CO)_4$. Perhaps the next most abundant is $Fe(CO)_5$. Other examples are $V(CO)_6$ and $Cr(CO)_6$. In some cases bonding favors compounds with two metal atoms per molecule, such as $(CO)_5Mn\text{-}Mn(CO)_5$ or $(CO)_4Co\text{-}Co(CO)_4$.

Organometallic Compounds Involving π-Electron Donors

Unsaturated hydrocarbons, such as ethylene, butadiene, cyclopentadiene, and benzene contain π-electrons that occupy orbitals that are not in a direct line between the two atoms bonded together but are above and below a plane through that line. These electrons can participate in bonds to metal atoms in organometallic compounds. Furthermore, the metal atoms in a number of organometallic compounds are bonded to both a π-electron donor organic species — most commonly the cyclopentadienyl anion with a -1 charge — and one or more CO molecules. A typical compound of this class is cyclopentadienylcobalt-dicarbonyl, $C_5H_5Co(CO)_2$. Examples of these compounds and of compounds consisting of metals bonded only to organic π-electron donors are shown in Figure 6.3.

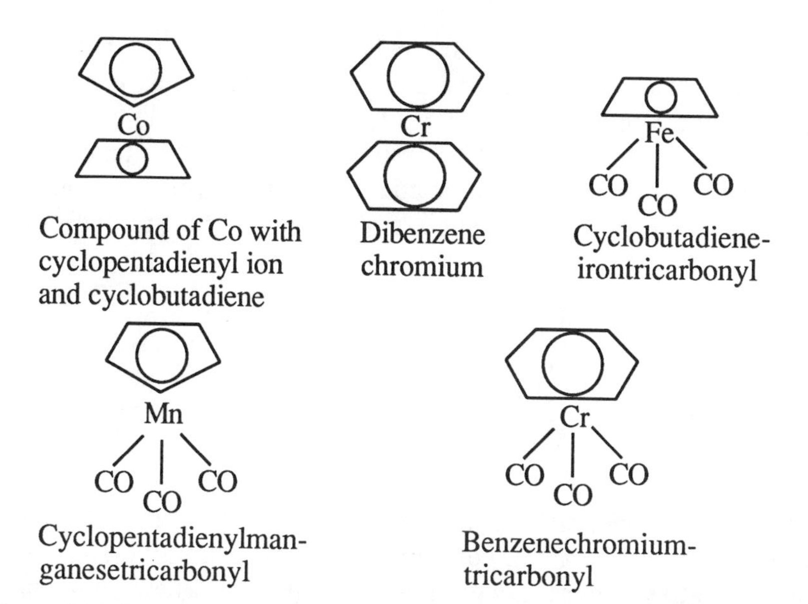

Figure 6.3. Compounds of metals with π-electron donor hydrocarbons and with carbon monoxide.

6.3. MIXED ORGANOMETALLIC COMPOUNDS

So far in this chapter the discussion has centered upon compounds in which all of the metal bonds are with carbon. A large number of compounds exist that have at least one bond between the metal and a C atom on an organic group, as well as other covalent or ionic bonds between the metal and atoms other than carbon. Because they have at least one metal-carbon bond, as well as properties, uses and toxicological effects typical of organometallic compounds, it is useful to consider such compounds along with organometallic compounds. Examples are monomethylmercury chloride, CH_3HgCl, in which the organometallic CH_3Hg^+ ion is ionically bonded to the chloride anion. Another example is phenyldichloroarsine, $C_6H_5AsCl_2$, in which a phenyl group is covalently bonded to arsenic through an As-C bond, and two Cl atoms are also covalently bonded to arsenic.

A number of compounds exist that consist of organic groups bonded to a metal atom through atoms other than carbon. Although they do not meet the strict definition thereof, such compounds can be classified as organometallics for the discussion of their toxicology and aspects of their chemistry. An example of such a compound is isopropyl titanate, $Ti(OC_3H_7)_4$, also called titanium isopropylate. This compound is a colorless liquid melting at 14.8°C and boiling at 104°C. Its behavior is more that of an organometallic compound than that of an inorganic compound, and by virtue of its titanium content it is not properly classified as an organic compound. The term "organometal" is sometimes applied to such a compound. For toxicological considerations it may be regarded as an organometallic compound.

$$Ti(OC_3H_7)_4$$

Isopropyl titanate

Several compounds are discussed in this chapter that have some organometallic character, but which also have formulas, structures, and properties of inorganic or organic compounds. These compounds could be called "mixed organometallics." However, so long as the differences are understood, compounds such as isopropyl titanate (see above) that do not meet all the criteria of organometallic compounds can be regarded as such for the discussion of their toxicities.

6.4. ORGANOMETALLIC COMPOUND TOXICITY

Some organometallic compounds have been known and used for decades, so that their toxicological properties are rather well known. Prominent among these are organoarsenicals used as drugs, organomercury fungicides, and tetramethyl- and tetraethyllead used as antiknock additives for gasoline. Since about 1950 there has been very substantial growth in chemical research devoted to organometallic compounds and large numbers and varieties of these compounds have been synthesized. Although the applications of organoarsenicals and organomercury compounds as human drugs and pesticides have declined sharply because of their toxicities, environmental effects, and the development of safer substitutes, a wide variety of new organometallic compounds has come into use for various purposes, such as catalysis and chemical synthesis. Toxicological experience is lacking for many of these compounds, so they should be treated with great caution until proven safe. Many are very reactive chemically, so that they are hazardous to directly exposed tissue, even if not toxic systemically.

6.5. COMPOUNDS OF GROUP 1A METALS

Lithium Compounds

Table 6.1 shows some organometallic lithium compounds. It is seen from their formulas that these compounds are ionic. As discussed in

Table 6.1. Some Organometallic Compounds of Lithium.

Name	Formula	Properties and Uses
Methyllithium	$Li^{+}\ {}^{-}CH_3$	Initiator for solution polymerization of elastomers
Ethyllithium	$Li^{+}\ {}^{-}CH_2CH_3$	Transparent crystals melting at 95°C, pyrophoric,[1] decomposes in water
Tert–butyllithium	$Li^{+}\ {}^{-}C(CH_3)_3$	Colorless crystalline solid subliming at 70–80°C, synthesis reagent
Phenyllithium	$Li^{+}\ {}^{-}C_6H_5$	Colorless pyrophoric solid used in Grignard-type reactions to attach a phenyl group

[1] Pyrophoric: Spontaneously flammable in air.

Section 6.2, 1A metals have low electronegativities and form ionic compounds with hydrocarbon anions. Of these elements, lithium tends to form metal-carbon bonds with the most covalent character; therefore, lithium compounds are more stable (though generally quite reactive) compared to other organometallic compounds of Group 1A metals, most likely to exist as liquids or low-melting-point solids, and generally more soluble in organic solvents.[3] These compounds are moisture-sensitive, both in the pure state and in solution, and can undergo spontaneous ignition when exposed to air.

The most widely used organolithium compound is *n*-butyllithium (see formulas of related compounds in Table 6.1), used as an initiator for the production of elastomers by solution polymerization, predominantly of styrene-butadiene.

Lithium forms a very unstable carbonyl, for which the toxicity is suspected of being high. The formula of this compound is LiCOCOLi, written in this manner to show that the two CO molecules form bridges between two Li atoms.

Unless otherwise known, the toxicities of lithium organometallic compounds should be regarded as those of lithium compounds and of organometallic compounds in general. The latter were discussed in Section 6.4. Lithium oxide and hydroxide are caustic bases, and they may be formed by the combustion of lithium organometallic compounds or by their reaction with water.

Lithium ion, Li^+, is a central nervous system toxicant that causes dizziness, prostration, anorexia, apathy, and nausea. It can also cause kidney damage and, in large doses, coma and death.

Compounds of Group 1A Metals Other than Lithium

As discussed in Section 6.2, Group 1A metals form ionic metal-carbon bonds. Organometallic compounds of Group 1A metals other than lithium have metal-carbon bonds with less of a covalent character than the corresponding bonds in lithium compounds and tend to be especially reactive. Compounds of rubidium and cesium are rarely encountered outside the laboratory, so their toxicological significance is relatively minor. Therefore, aside from lithium compounds, the toxicology of sodium and potassium compounds is of most concern.

Both sodium and potassium salts are natural constituents of body tissues and fluids as Na^+ and K^+ ions, respectively, and are not themselves toxic at normal physiological levels. The oxides and hydroxides of both these metals are very caustic, corrosive substances that damage exposed tissue. Oxides are formed by the combustion of sodium and potassium organometallics, and hydroxides are produced by the reaction of the oxides with water or by direct reaction of the organometallics with water, as shown below for cyclopentadienylsodium:

$$C_5H_5^-Na^+ + H_2O \rightarrow C_5H_6 + NaOH \quad (6.5)$$

Both sodium and potassium form carbonyl compounds, NaCO and $(KCO)_6$, respectively. Both compounds are highly reactive solids prone to explode when exposed to water or air. Decomposition of the carbonyls gives off caustic oxides and hydroxides of Na and K, as well as toxic carbon monoxide.

Sodium and potassium form alkoxide compounds with the general formula $M^{+-}OR$, in which R is a hydrocarbon group. Typically, sodium reacts with methanol:

$$2CH_3OH + 2Na \rightarrow 2Na^{+}\,{}^{-}OCH_3 + H_2 \qquad (6.6)$$

to yield sodium methoxide and hydrogen gas. The alkoxide compounds are highly basic and caustic, reacting with water to form the corresponding hydroxides as illustrated by the following reaction:

$$K^{+}\,{}^{-}OCH_3 + H_2O \rightarrow KOH + CH_3OH \qquad (6.7)$$

6.6. COMPOUNDS OF GROUP 2A METALS

The organometallic compound chemistry of the 2A metals is similar to that of the 1A metals and ionically-bonded compounds predominate. As is the case with lithium in Group 1A, the first 2A element, beryllium, behaves atypically, with a greater covalent character in its metal-carbon bonds.

Beryllium organometallic compounds should be accorded the respect due all beryllium compounds because of the extreme toxicity of beryllium (see Section 5.4). Dimethylberyllium, $Be(CH_3)_2$, is a white solid having needle-like crystals. When heated to decomposition, it gives off highly toxic beryllium oxide fumes. Diethylberyllium, $Be(C_2H_5)_2$, with a melting point of 12˚C and a boiling point of 110˚C, is a colorless liquid at room temperature and is especially dangerous because of its volatility.

Magnesium

The organometallic chemistry of magnesium has been of the utmost importance for many decades because of **Grignard reagents**, the first of which was made by Victor Grignard around 1900 by the reaction:

$$\underset{\text{Iodomethane}}{\text{H}-\overset{\text{H}}{\underset{\text{H}}{\text{C}}}-\text{I}} + \text{Mg} \rightarrow \underset{\text{Methylmagnesium iodide}}{\text{H}-\overset{\text{H}}{\underset{\text{H}}{\text{C}}}\text{-Mg}^{+}\text{I}^{-}} \quad (6.8)$$

Grignard reagents are particularly useful in organic chemical synthesis for the attachment of their organic component ($-CH_3$ in the preceding example) to another organic molecule. The development of Grignard reagents was such an advance in organic chemical synthesis that in 1912 Victor Grignard received the Nobel Prize for his work.

Grignard reagents can cause damage to skin or pulmonary tissue in the unlikely event that they are inhaled. These reagents react rapidly with both water and oxygen, releasing a great deal of heat in the process. Ethyl ether solutions of methylmagnesium bromide (CH_3MgBr) are particularly hazardous because of the spontaneous ignition of the reagent and the solvent ether in which it is contained when the mixture contacts water, such as water on a moist laboratory bench top.

The simplest dialkyl magnesium compounds are dimethylmagnesium, $Mg(CH_3)_2$, and diethylmagnesium, $Mg(C_2H_5)_2$. Both are pyrophoric compounds that are violently reactive to water and steam and that self-ignite in air, the latter even in carbon dioxide (like the elemental form, magnesium in an organometallic compound removes O from CO_2 to form MgO and release elemental carbon). Diethylmagnesium has a melting point of 0°C and is a liquid at room temperature. Diphenylmagnesium, $Mg(C_6H_5)_2$, is a feathery solid, somewhat less hazardous than the dimethyl and diethyl compounds. It is violently reactive with water and is spontaneously flammable in humid air, but not dry air.

Unlike the caustic oxides and hydroxides of Group IA metals, magnesium hydroxide ($Mg(OH)_2$), formed by the reaction of air and water with magnesium organometallic compounds, is a relatively benign substance that is used as a food additive and ingredient of milk of magnesia.

Calcium, Strontium, and Barium

It is much more difficult to make organometallic compounds of Ca, Sr, and Ba than it is to make those of the first two Group 2A metals. Whereas organometallic compounds of beryllium and magnesium have metal-carbon bonds with a significant degree of covalent character, the Ca/Sr/Ba organometallic compounds are much more ionic. These compounds are extremely reactive to water, water vapor, and atmospheric oxygen. There are relatively few organometallic compounds of calcium, strontium, and barium; their industrial uses are few, so their toxicology is of limited concern. Grignard reagents in which the metal is calcium rather than magnesium (general formula RCa^+X^-) have been prepared, but are not as useful for synthesis as the corresponding magnesium compounds.

6.7. COMPOUNDS OF GROUP 2B METALS

It is convenient to consider the organometallic compound chemistry of the Group 2B metals immediately following that of the 2A metals because both have two 2*s* electrons and no partially filled *d* orbitals. The group 2B metals — zinc, cadmium, and mercury — form an abundance of organometallic compounds, many of which have significant uses. Furthermore, cadmium and mercury (both discussed in Chapter 5) are notably toxic elements, so the toxicological aspects of their organometallic compounds is of particular concern. Therefore, the organometallic compound chemistry of each of the 2B metals will be discussed separately.

Zinc

A typical synthesis of a zinc organometallic compound is given by the reaction below in which the Grignard-type compound CH_3ZnI is an intermediate:

$$\begin{array}{c}\text{H}\\|\\\text{H—C—I}\\|\\\text{H}\end{array} + 2\text{Zn} \rightarrow \begin{array}{c}\text{H}\quad\ \ \text{H}\\|\qquad\ |\\\text{H—C-Zn-C—H}\\|\qquad\ |\\\text{H}\quad\ \ \text{H}\end{array} + \text{ZnI}_2 \qquad (6.9)$$

Dimethylzinc has a rather low melting temperature of -40°C and it boils at 46°C. At room temperature it is a mobile, volatile liquid that undergoes self-ignition in air and reacts violently with water. The same properties are exhibited by diethylzinc, $(C_2H_5)Zn$, which melts at -28°C and boils at 118°C. Diphenylzinc, $(C_6H_5)Zn$, is considerably less reactive than its methyl and ethyl analogs; it is a white crystalline solid melting at 107°C. Zinc organometallics are similar in many respects to their analogous magnesium compounds (see Section 6.6), but do not react with carbon dioxide, as do some of the more reactive magnesium compounds. An example of an organozinc compound involving a π-bonded group is that of methylcyclopentadienylzinc, shown in Figure 6.4.

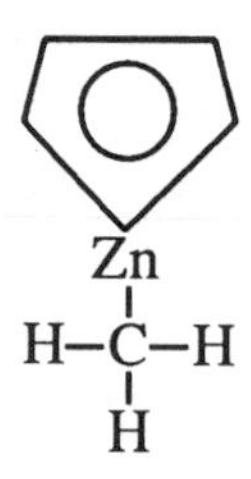

Figure 6.4. Methylcyclopentadienylzinc. The monomer shown exists in the vapor phase. In the solid phase a polymeric form exists.

Zinc organometallic compounds should be accorded the same caution in respect to toxicology as that of organometallic compounds in general. The combustion of highly flammable organozinc compounds such as dimethyl and diethyl compounds produces very finely divided particles of zinc oxide fumes as illustrated by the reaction:

$$2(CH_3)_2Zn + 8O_2 \rightarrow 2ZnO + 4CO_2 + 6H_2O \quad (6.10)$$

Although zinc oxide is used as a healing agent and food additive, inhalation of zinc oxide fume particles causes zinc **metal fume fever** characterized by elevated temperature and "chills." The toxic effect of zinc fume has been attributed to its flocculation in lung airways, which prevents maximum penetration of air to the alveoli and perhaps activates endogenous pyrogen in blood leukocytes.[4] An interesting aspect of this discomfiting but less-than-deadly affliction is the immunity that exposed individuals develop to it, but which is lost after only a day or two of non-exposure. Thus workers exposed to zinc fume usually suffer most from the metal fume fever at the beginning of the work week, and less with consecutive days of exposure as their systems adapt to the metal fume.

Diphenylzinc illustrates the toxicity hazard that may obtain from the organic part of an organometallic compound upon decomposition. Under some conditions this compound can react to release toxic phenol (see Chapter 9):

$$C_6H_5-Zn-C_6H_5 \xrightarrow[\{O_2\}]{H_2O} C_6H_5-OH + \text{Zinc species} \quad (6.11)$$

A number of zinc compounds with organic constituents (e.g., zinc salts of organic acids) have therapeutic uses. These include anti-dandruff zinc pyridinethione, antifungal zinc undecylenate used to treat athlete's foot, zinc stearate and palmitate (zinc soap), and antibacterial zinc bacitracin. Zinc naphthenate is used as a low-toxicity wood preservative and zinc phenolsulfonate has insecticidal properties, and was once used as an intestinal antiseptic. The inhalation of zinc soaps by infants has been known to cause acute fatal pneumonitis[5] characterized by lung lesions similar to, but more serious than, those caused by talc. Zinc pyridine thione (zinc 2-pyridinethiol-1-oxide) has been shown to cause retinal detachment and blindness in dogs;[6] this effect is an apparently species-specific effect because laboratory tests at the same and even much higher dosages in monkeys and rodents do not show the same effect.

Cadmium

In the absence of water, cadmium halides, CdX_2, react with organolithium compounds as shown by the following example:

$$CdBr_2 + 2Li^{+}\,{}^{-}C_6H_5 \rightarrow 2LiBr + C_6H_5-Cd-C_6H_5 \quad (6.12)$$

Dimethylcadmium, $(CH_3)_2Cd$, is an oily liquid at room temperature and has a very unpleasant odor. The compound melts at -4.5°C and boils at 106°C. It decomposes in contact with water. Diethylcadmium is likewise an oil; it melts at -21°C and boils at 64°C and reacts explosively with oxygen in air. Dipropylcadmium, $(C_3H_7)_2Cd$, is an oil that melts at -83°C, boils at 84°C, and reacts with water. The dialkyl cadmium compounds are distillable, but decompose above about 150°C, evolving toxic cadmium fume.

The toxicology of cadmium organometallic compounds is of particular concern because of the high toxicity of cadmium. The organometallic compounds of cadmium form vapors that can be inhaled and that can cross membranes because of their lipid solubility. The reaction of cadmium organometallic compounds with water can release highly toxic fumes of cadmium and CdO. Inhalation of these fumes can cause chronic cadmium poisoning and death. The toxicological aspects of cadmium are discussed in Section 5.4.

Mercury

In 1853 E. Frankland made the first synthetic organomercury compound by the photochemical reaction

$$2Hg + 2CH_3I + h\nu \rightarrow (CH_3)_2Hg + HgI_2 \qquad (6.13)$$

where sunlight was used as the light energy source ($h\nu$). Numerous synthetic routes are now available for the preparation of a variety of mercury organometallic compounds.

In the late 1800s and early 1900s numerous organomercury pharmaceutical compounds were synthesized and used. These have since been replaced by more effective and safe non-mercury substitutes. Organomercury compounds have been widely used as pesticidal fungicides (see Figure 6.5), but these applications are now declining because of the adverse effects of mercury in the environment.

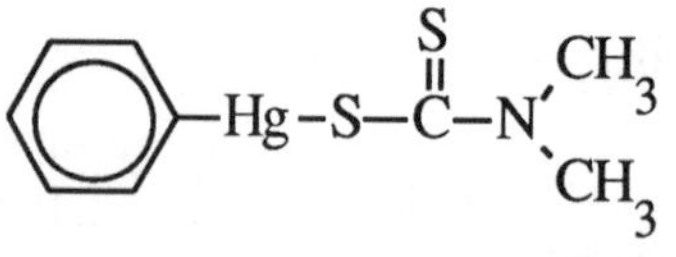

$$C_6H_5\text{-}Hg\text{-}S\text{-}C(=S)\text{-}N(CH_3)_2$$

Phenylmercurydimethyldithio-carbamate (slimicide for wood pulp and mold retardent for paper)

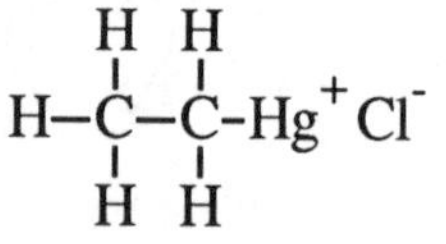

$$H\text{-}\underset{H}{\overset{H}{C}}\text{-}\underset{H}{\overset{H}{C}}\text{-}Hg^+\,Cl^-$$

Ethylmercury chloride (seed fungicide)

Figure 6.5. Two organomercury compounds that have been used for fungicidal purposes.

The most notorious mercury compounds in the environment are monomethylmercury (CH_3Hg^+) salts and dimethylmercury ($(CH_3)_2Hg$). The latter compound is both soluble and volatile and the salts of the monomethylmercury cation are soluble. These compounds are produced from inorganic mercury in sediments by anaerobic bacteria through the action of methylcobalamin, a vitamin B12 analog and intermediate in the synthesis of methane:

$$HgCl_2(s) \xrightarrow{\text{Methylcobalamin}} CH_3Hg^+(aq) + 2Cl^- \quad (6.14)$$

The preceding reaction is favored in somewhat acidic water in which anaerobic decay, which often produces CH_4, is occurring. If the water is neutral or slightly alkaline, dimethylmercury formation is favored; this volatile compound may escape to the atmosphere. Discovered around 1970, the biosynthesis of the methylmercury species in sediments was an unpleasant surprise, in that it provides a means for otherwise insoluble inorganic mercury compounds to get into natural waters. Furthermore, these species are lipid-soluble, so that they undergo bioaccumulation and biomagnification in aquatic organisms. Fish tissue often contains more than 1000 times the concentration of mercury as does the surrounding water.

The toxicity of mercury is discussed in Section 5.4. Some special considerations apply to organomercury compounds, the foremost of which is their lipid solubility and resulting high degree of absorption and facile distribution through biological systems. The lipid solubilities and high vapor pressures of the methylmercuries favor their absorption by the pulmonary route. These compounds also can be absorbed through the skin, and their uptake approaches 100% (compared to less than 10% for inorganic mercury compounds) in the gastrointestinal tract.

With respect to distribution in the body, the methylmercury species behave more like mercury metal, Hg(0), than inorganic mercury(II), Hg^{2+}. Like elemental mercury, methylmercury compounds traverse the blood-brain barrier and affect the central nervous system. However, the psychopathological effects of methylmercury compounds (laughing, crying, impaired intellectual abilities) are different from those of elemental mercury (irritability, shyness).

Mono- and diphenylmercury have toxicological effects much like those of inorganic mercury(II) because of their rapid hydrolysis in the body:

$$C_6H_5{-}Hg{-}C_6H_5 + 2H_2O \rightarrow Hg(OH)_2 + 2\,C_6H_6 \xrightarrow{\{O\}} 2\,C_6H_5{-}OH \quad (6.15)$$

Phenol

6.8. ORGANOTIN COMPOUNDS

Global production of organotin compounds is on the order of 40,000 metric tons per year and consumes about 7–8 percent of the tin used each year. Of all the metals, tin has the greatest number of organometallic compounds in commercial use.[6] Major industrial uses include applications of tin compounds in fungicides, acaricides, disinfectants, antifouling paints, stabilizers to lessen the effects of heat and light in PVC plastics, catalysts, and precursors for the formation of films of SnO_2 on glass. Tributyl tin chloride and related tributyl tin (TBT) compounds have bactericidal, fungicidal, and insecticidal properties and are of particular environmental significance because of growing use as industrial biocides. In addition to tributyl tin chloride, other tributyl tin compounds used as biocides include the hydroxide, the naphthenate, bis(tributyltin) oxide, and tris(tributylstannyl) phosphate. A major use of TBT is in boat and ship hull coatings to prevent the growth of fouling organisms.[7] Other applications include preservation of wood, leather, paper, and textiles.[8] Because of their antifungal activity TBT compounds are used as slimicides in cooling tower water.

In addition to synthetic organotin compounds, methylated tin species can be produced biologically in the environment. Figure 6.6 gives some examples of the many known organotin compounds.

Toxicology of Organotin Compounds

Many organotin compounds have the general formula, R_nSnX_{4-n}, where R is a hydrocarbon group and X is an inorganic entity, such as a chlorine atom, or an organic group bonded to tin through a non-carbon atom (for example, acetate bonded to Sn through an O atom).[8] As a general rule, in a series of these compounds, toxicity is at a maximum value for n = 3. Furthermore, the toxicity is generally more dependent upon the nature of the R groups than upon X.

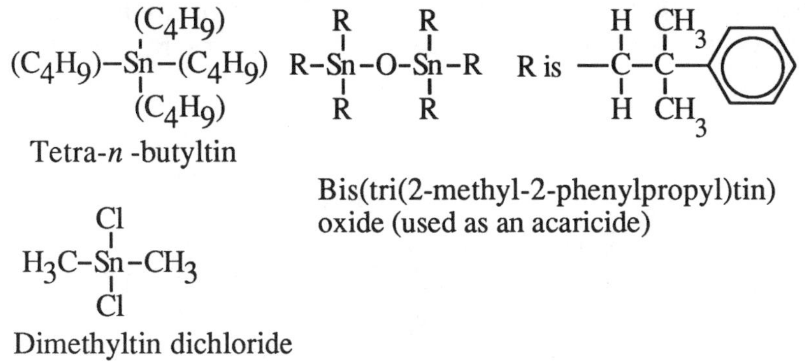

$(C_4H_9)_4Sn$

Tetra-*n* -butyltin

$R_3Sn{-}O{-}SnR_3$, R is $-CH_2-C(CH_3)_2-C_6H_5$

Bis(tri(2-methyl-2-phenylpropyl)tin) oxide (used as an acaricide)

$(CH_3)_2SnCl_2$

Dimethyltin dichloride

Figure 6.6. Examples of organotin compounds.

Organotin compounds are readily absorbed through the skin, and skin rashes may result. Organotin compounds, especially those of the R_3SnX type, bind to proteins, probably through the sulfur on cysteine and histidine residues. Interference with mictochondrial function by several mechanisms appears to be the mode of biochemical action leading to toxic responses.

6.9. ORGANOLEAD COMPOUNDS

The toxicities and environmental effects of organolead compounds are particularly noteworthy because of the widespread use and distribution of tetraethyllead as a gasoline additive (see structure in Figure 6.2). Although more than 1000 organolead compounds have been synthesized, those of commercial and toxicological importance are largely limited to the alkyl (methyl and ethyl) compounds and their salts, examples of which are shown in Figure 6.7.

In addition to manufactured organolead compounds, the possibility exists of biological methylation of lead, such as occurs with mercury (see Section 6.7). However, there is a great deal of uncertainty regarding biological methylation of lead in the environment.

Toxicology of Organolead Compounds

Because of the large amounts of tetraethyllead used as a gasoline additive, the toxicology of this compound has been investigated much more extensively than that of other organolead compounds and is discussed briefly here. Tetraethyllead is a colorless, oily liquid with a strong affinity for lipids and is considered highly toxic by inhalation,

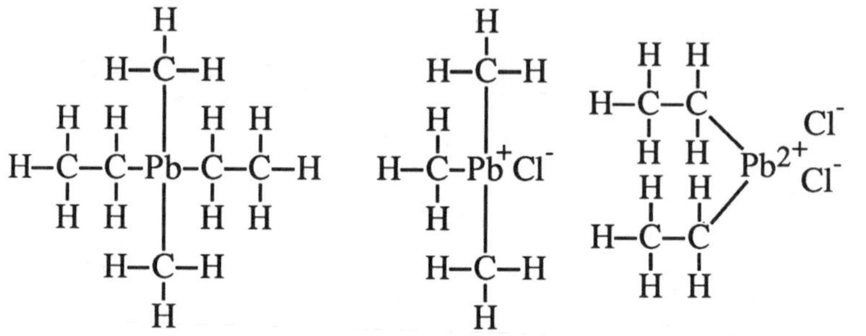

Dimethyldiethyllead — Trimethyllead chloride — Diethyllead dichloride

Figure 6.7. Alkyllead compounds and salts.

ingestion, and absorption through the skin. The toxicological action of tetraethyllead is different from that of inorganic lead. As one manifestation of this difference, chelation therapy is ineffective for the treatment of tetraethyllead poisoning. Symptoms of tetraethyllead poisoning reflect effects upon the central nervous system. Among these symptoms are fatigue, weakness, restlessness, ataxia, psychosis, and convulsions. In cases of fatal tetraethyllead poisoning, death has occurred as soon as one or two days after exposure. Fatalities have been comparatively rare, considering the widespread use of tetraethyllead. Recovery from poisoning by this compound tends to be slow. Its toxic action appears to involve its metabolic conversion to the triethyl form.

6.10. ORGANOARSENIC COMPOUNDS

There are two major sources of organoarsenic compounds — those produced for commercial applications and those produced from the biomethylation of inorganic arsenic by microorganisms. Many different organoarsenic compounds have been identified.

Organoarsenic Compounds from Biological Processes

The reactions that follow illustrate the production of organoarsenic compounds by bacteria.[6] In a reducing environment, arsenic(V) is reduced to arsenic(III):

$$H_3AsO_4 + 2H^+ + 2e^- \rightarrow H_3AsO_3 + H_2O \qquad (6.16)$$

Through the action of methylcobalamin in bacteria, arsenic(III) is methylated to methyl- then to dimethylarsinic acid:

$$H_3AsO_3 \rightarrow \begin{array}{ccccc} & H & & O & \\ & | & & \| & \\ H- & C & - & As & -OH \\ & | & & | & \\ & H & & OH & \end{array} \qquad (6.17)$$

$$\begin{array}{ccccc} & H & & O & \\ & | & & \| & \\ H- & C & - & As & -OH \\ & | & & | & \\ & H & & OH & \end{array} \rightarrow \begin{array}{ccccccc} & H & & O & & H & \\ & | & & \| & & | & \\ H- & C & - & As & - & C & -H \\ & | & & | & & | & \\ & H & & OH & & H & \end{array} \qquad (6.18)$$

Dimethylarsinic acid can be reduced to volatile dimethylarsine:

$$\begin{array}{ccccccc} & H & & O & & H & \\ & | & & \| & & | & \\ H- & C & - & As & - & C & -H \\ & | & & | & & | & \\ & H & & OH & & H & \end{array} + 4H^+ + 4e^- \rightarrow \begin{array}{ccccccc} & H & & H & & H & \\ & | & & | & & | & \\ H- & C & - & As & - & C & -H \\ & | & & & & | & \\ & H & & & & H & \end{array} + 2H_2O \qquad (6.19)$$

Methylarsinic acid and dimethyarsinic acid are the two organoarsenic compounds that are most likely to be encountered in the environment.

Biomethylated arsenic was responsible for numerous cases of arsenic poisoning in Europe during the 1800s. Under humid conditions, arsenic in plaster and wallpaper pigments was converted to biomethylated forms, as manifested by the strong garlic odor of the products, and people sleeping and working in the rooms became ill from inhaling the volatile organoarsenic compounds.

Synthetic Organoarsenic Compounds

Although now essentially obsolete for the treatment of human diseases because of their toxicities, organoarsenic compounds were the first synthetic organic pharmaceutical agents and were widely used in the early 1900s. The first pharmaceutical application was that of atoxyl (the sodium salt of 4-aminophenylarsinic acid), which was used to treat sleeping sickness. The synthesis of Salvarsan by Dr. Paul Ehrlich in 1907 was a development that may be considered the beginning of modern **chemotherapy** (chemical treatment of disease). Salvarsan was widely used for the treatment of syphilis.

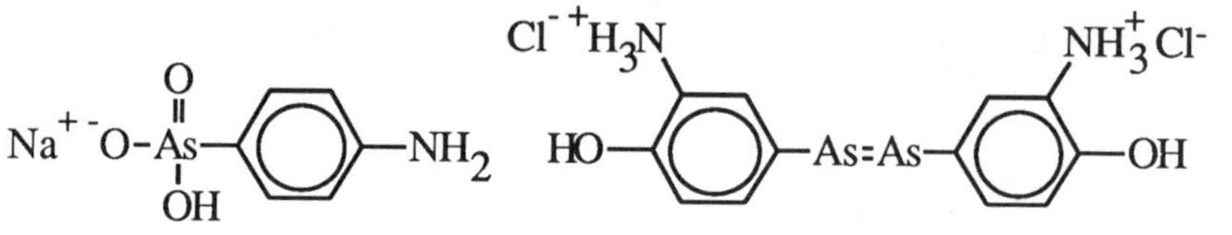

Atoxyl

Salvarsan

Organoarsenic compounds are used as animal feed additives. The major organoarsenic feed additives and their uses are summarized in Figure 6.8.

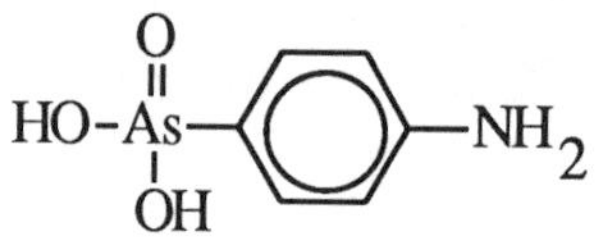

Arsanilic acid

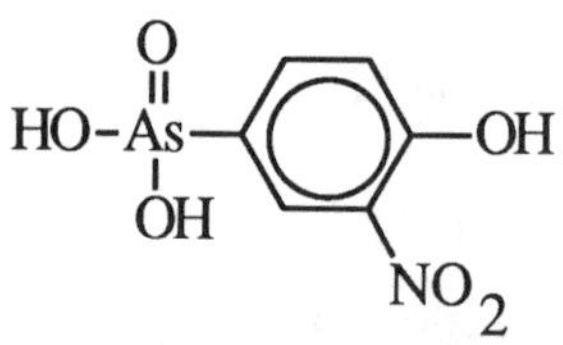

3-Nitro-4-hydroxyphenyl-arsinic acid (Roxarsone)

N-carbamoylarsinic acid (Carbarsone)

Figure 6.8. Major organoarsenic animal feed additives. Arsanilic acid and Roxarsone are used to control swine dysentery and increase the rate of gain relative to the amount of feed in swine and chickens. Carbarsone and nitarsone (4-nitrophenylarsanilic acid) act as antihistomonads in chickens.

Toxicities of Organoarsenic Compounds

The toxicities of organoarsenic compounds vary over a wide range. In general, the toxicities are less for those compounds that are not metabolized in the body and that are excreted in an unchanged form. Examples of such compounds are the animal feed additives shown in Figure 6.8. Metabolic breakdown of organoarsenic compounds to inorganic forms is correlated with high toxicity. This is especially true when the product is inorganic arsenic(III), which, for the most part, is more toxic than arsenic(V). The toxicity of arsenic(III) is related to its strong affinity for sulfhydryl (–SH) groups. Detrimental effects are especially likely to occur when sulfhydryl groups are adjacent to each other on the active sites of enzymes, enabling chelation of the arsenic and inhibition of the enzyme.

To a certain extent toxic effects of dimethylarsinic acid (cacodylic acid) have been observed because of its applications as a herbicide and former uses of its sodium salt for the treatment of human skin disease and leukemia. It is most toxic via ingestion because the acidic medium in the stomach converts the compound to inorganic arsenic(III). A portion of inorganic arsenic in the body is converted to dimethylarsinic acid, which is excreted in urine, sweat, and exhaled air, accompanied by a strong garlic odor. Roxarsone has a relatively high acute toxicity to rats and dogs.[9] Among the effects observed in these animals are internal hemorrhage, kidney congestion, and gastroenteritis. Rats fed fatal doses of about 400 ppm in the diet exhibited progressive weakness prior to death.

6.11. ORGANOSELENIUM AND ORGANOTELLURIUM COMPOUNDS

Organo compounds of the two Group 6A elements, selenium and tellurium, are of considerable environmental and toxicological importance. Organoselenium and organotellurium compounds are produced both synthetically and by microorganisms. The selenium compounds are the more significant because of the greater abundance of this element.

Organoselenium Compounds

The structures of three common organoselenium compounds produced by organisms are given in Figure 6.9.

Some organisms convert inorganic selenium to dimethylselenide. Several genera of fungi are especially adept at this biomethylation process, and their activities are readily detected from the very strong "ultragarlic" odor of the product. The bioconversion of inorganic selenium(II) and selenium(VI) to dimethylselenide and dimethyldiselenide occurs in animals such as rats, and the volatile compounds are evolved with exhaled air. Another organoselenium compound produced by bacteria is dimethylselenone. Some synthetic organoselenium compounds have selenium as part of a ring, such as is the case with the cyclic ether, 1,4-diselenane.

```
   H     H             H           H           H  O  H
   |     |             |           |           |  ||  |
H–C–Se–C–H         H–C–Se–Se–C–H        H–C–Se–C–H
   |     |             |           |           |  ||  |
   H     H             H           H           H  O  H
```

Dimethylselenide   Dimethyldiselenide   Dimethylselenone

Figure 6.9. Example organoselenium compounds.

Inorganic selenium compounds are rather toxic, and probably attach to protein sulfhydryl groups, much like inorganic arsenic. In general, organoselenium compounds are regarded as being less toxic than inorganic selenium compounds.

Organotellurium Compounds

Inorganic tellurium is used in some specialized alloys, to color glass, and as a pigment in some porcelain products. The breath of workers exposed to inorganic tellurium has a garlic odor, perhaps indicative of bioconversion to organotellurium species. Dimethyltelluride can be produced by fungi from inorganic tellurium compounds. Tellurium is a rather rare element in the geosphere and in water, so that biomethylation of this element is unlikely to be a major environmental problem. In general, the toxicities of tellurium compounds are less than those of their selenium analogs.

LITERATURE CITED

1. Haiduc, Ionel, and J. J. Zuckerman, *Basic Organometallic Chemistry*, Walter de Gruyter, Berlin/New York, 1985.

2. Lukehart, Charles M., *Fundamental Transition Metal Organometallic Chemistry*, Brooks/Cole Publishing Co., Monterey, CA, 1985.

3. Bach, Ricardo, R. B. Ellestad, C. W. Camienski, and J. R. Wasson, "Lithium and Lithium Compounds," in *Kirk-Othmer Concise Encyclopedia of Chemical Technology,* Wiley-Interscience, New York, 1985, pp. 706-707.

4. Hammond, Paul B., and Robert Beliles, "Metals," Chapter 17 in *Casarett and Doull's Toxicology,* 2nd ed., John Doull, Curtis D. Klaassen, and Mary O. Amdur, Eds., Macmillan Publishing Co., New York, 1980, pp. 409-467.

5. "Zinc Stearate," in *Clinical Toxicology of Commercial Products,* Robert E. Gosselin, Roger P. Smith, and Harold C. Hodge, Williams and Wilkins, Baltimore, 1984.

6. Craig, P. J., Ed., *Organometallic Compounds in the Environment,* John Wiley and Sons, New York, 1986.

7. Seligman, Peter F., et al, "Distribution and Fate of Tributyltin in the Marine Environment," *American Chemical Society Division of Environmental Chemistry Preprint Extended Abstracts*, **28**, 573–579 (1988).

8. Clark, Elizabeth M., Robert M. Sterritt, and John N. Lester, "The Fate of Tributyltin in the Aquatic Environment," *Environmental Science and Technology*, **22**, 600–604 (1988).

9. Gosselin, Robert E., Roger P. Smith, and Harold C. Hodge, "4-Hydroxy-3-nitrophenylarsonic Acid," in *Clinical Toxicology of Commercial Products*, 5th ed., Williams and Wilkins, Baltimore/London, 1984, p. II-132.

7

Toxic Inorganic Compounds

7.1. INTRODUCTION

In Chapter 5 elements were discussed that as a rule tend to be toxic in their various forms. Chapter 7 covers toxic inorganic compounds of elements that are not themselves generally regarded as being toxic. These elements include for the most part the lighter nonmetals located in the upper right of the periodic table (Figure 2.1), and exclude the heavy metals. Most of the elements involved in the inorganic compounds discussed in this chapter are those that are essential for life processes. Any division between "toxic" and "nontoxic" elements is by nature artificial in that most of the heavy metals have compounds of relatively low toxicity, and there are deadly compounds that contain elements essential for life. The toxicities of inorganic compounds are covered in detail in a reference work on the subject.[1]

Chapter Organization

In general this chapter is organized in the order of increasing atomic number of the elements that are covered. Inorganic compounds of carbon, atomic number 6, are discussed first, followed by toxic inorganic compounds of nitrogen, atomic number 7. The next element, oxygen, occurs in so many different inorganic compounds that it is not discussed in a separate category except for its toxic elemental form, ozone. The halogens — fluorine, chlorine, bromine, and iodine — are discussed as a group because of their chemical similarities. The other major elements whose toxic inorganic compounds are discussed are silicon, phosphorus, and sulfur.

7.2. TOXIC INORGANIC CARBON COMPOUNDS

Cyanide

Cyanide, either in the form of gaseous **hydrogen cyanide** (HCN) or **cyanide ion** (CN^- present in cyanide salts such as KCN) is a notably toxic substance. Cyanide is a rapidly acting poison[2] and the fatal oral dose to humans is believed to be only 60–90 mg. Hydrogen cyanide and cyanide salts have numerous uses, so that exposure is certainly possible. Hydrogen cyanide is used as a fumigant to kill pests such as rodents in warehouses, grain storage bins, greenhouses, and holds of ships, where its high toxicity and ability to penetrate obscure spaces are advantageous. Cyanide salt solutions are used to extract some metals such as gold from ores, in metal refining, in metal plating, and for salvaging silver from exposed photographic and X-ray film. Cyanide is used in various chemical syntheses. Polyacrylic polymers may evolve HCN during combustion, adding to the toxic gases that are usually responsible for deaths in fires.

Biochemical Action of Cyanide

Cyanide inhibits an enzyme (see enzyme inhibition, Section 4.4) involved in a key step in the oxidative phosphorylation pathway by which the body utilizes oxygen in cell mitochondria. This enzyme is ferricytochrome oxidase, an iron-containing metalloprotein. Cyanide bonds to the iron(III) of the enzyme, preventing its reduction to iron(II). The result is that ferrouscytochrome oxidase, which is required to react with O_2, is not formed and utilization of oxygen in cells is prevented, leading to rapid cessation of metabolic processes.

Carbon Monoxide

Carbon monoxide, CO, is a toxic industrial gas produced by the incomplete combustion of carbonaceous fuels. It is used as a reductant for metal ores, for chemical synthesis, and as a fuel. As an environmental toxicant (see Section 1.2) it is responsible for a significant number of accidental poisonings annually. Observable acute effects of carbon monoxide exposure in humans cover a wide range of symptoms and severity. These include impairment of judgment and visual perception at CO levels of 10 parts per million (ppm) in air; dizziness, headache, weariness (100 ppm); loss of consciousness (250 ppm); and rapid death (1,000 ppm). Chronic effects of long-term low-level exposure to carbon monoxide are not well known, but are suspected of including disorders of the respiratory system and the heart.[3]

Biochemical Action of Carbon Monoxide

Carbon monoxide enters the bloodstream through the lungs and reacts with hemoglobin (Hb) as follows where O_2Hb is oxyhemoglobin and COHb is carboxyhemoglobin:

$$O_2Hb + CO \rightarrow COHb + O_2 \quad (7.1)$$

Carboxyhemoglobin is several times more stable than oxyhemoglobin and ties up the hemoglobin so that it cannot reach body tissues.

Cyanogen, Cyanamide, and Cyanates

Cyanogen, NCCN, is a colorless violently flammable gas with a pungent odor. It may cause permanent injury or even death in exposed individuals. Fumes produced by the reaction of cyanogen with water or acids are highly toxic.

Cyanamide (shown here):

$$\begin{matrix} H \\ & \backslash \\ & & N-C\equiv N \\ & / \\ H \end{matrix}$$

and calcium cyanamide (CaNCN) are used as fertilizers and raw materials. Calcium cyanamide is employed for the desulfurization and nitridation of steel. Inhalation or oral ingestion of cyanamide causes dizziness, lowers blood pressure, and increases rates of pulse and respiration.[4] Calcium cyanamide acts as a primary irritant to the skin and to nose and throat tissues.

Cyanic acid, HOCN, (bp 23.3°C, mp -86°C) is a dangerously explosive liquid with an acrid odor. The acid forms cyanate salts, such as NaOCN and KOCN. During decomposition from heat or contact with strong acid, cyanic acid evolves very toxic fumes.

7.3. NITROGEN OXIDES

The two most common oxides of nitrogen are **nitric oxide** (NO) and **nitrogen dioxide** (NO_2), designated collectively as NO_x. Nitric oxide is produced in combustion processes from organically bound nitrogen endogenous to fossil fuels (particularly coal, heavy fuel oil, and shale oil) and from atmospheric nitrogen under the conditions that exist in an internal combustion engine as shown by the two following reactions:

$$2N(\text{fossil fuel}) + O_2 \rightarrow 2NO \quad (7.2)$$

$$N_2 + O_2 \xrightarrow[\text{engine}]{\text{Internal combustion}} 2NO \quad (7.3)$$

Under the conditions of photochemical smog formation[5] nitric oxide is converted to nitrogen dioxide by the following overall reaction:

$$2NO + O_2 \xrightarrow[\text{ical processes}]{\text{Organics, photochem-}} 2NO_2 \quad (7.4)$$

This conversion consists of complex chain reactions involving light energy and unstable reactive intermediate species. The conditions required are stagnant air, low humidity, intense sunlight, and the presence of reactive hydrocarbons, particularly those from automobile exhausts. Of the NO_x constituents, NO_2 is generally regarded as the more toxic, although all nitrogen oxides and potential sources thereof (such as nitric acid in the presence of oxidizable organic matter) should be accorded the same respect as nitrogen dioxide.

Effects of NO_2 Poisoning

Inhalation of NO_2 causes severe irritation of the innermost parts of the lungs resulting in pulmonary edema and fatal bronchiolitis fibrosa obliterans. Inhalation for even very brief periods of time of air containing 200–700 ppm of NO_2 can be fatal. The biochemical action of NO_2 includes disruption of some enzyme systems, such as lactic dehydrogenase. Nitrogen dioxide probably acts as an oxidizing agent similar to, though weaker than, ozone, which is discussed in Section 7.4. Included is the formation of free radicals, particularly the hydroxyl radical HO·. Like ozone, it is likely that NO_2 causes

lipid peroxidation. This is a process in which the C=C double bonds in unsaturated lipids are attacked by free radicals and undergo chain reactions in the presence of O_2, resulting in their oxidative destruction.[6]

Nitrous Oxide

Nitrous oxide, once commonly known as "laughing gas," is used as an oxidant gas and in dental surgery as a general anesthetic. It is a central nervous system depressant and can act as an asphyxiant.

7.4. OZONE

Ozone (O_3) is a reactive and toxic form of elemental oxygen. It is used as an oxidant for chemical synthesis and for the disinfection of water. In the latter application it avoids the production of potentially toxic organochlorine by-products. Ozone is also used for the destruction of organic compounds responsible for odors from municipal wastewater treatment plants and industrial operations. For these purposes it is produced by an electrical discharge through dry air. The production of pollutant atmospheric ozone occurs under conditions of photochemical smog formation as discussed in the preceding section. Ozone is also produced from ultraviolet light passing through air by the reactions:

$$O_2 + h\nu \rightleftarrows O + O \tag{7.5}$$

$$O + O_2 + M \rightleftarrows O_3 + M \tag{7.6}$$

where $h\nu$ represents the energy of a photon of ultraviolet radiation and M is an energy-absorbing third body, usually a molecule of O_2 or N_2. The odor of ozone produced by any of these reactions can be detected around inadequately vented instruments, such as spectrofluorometers, that have intense ultraviolet sources.

Toxicity of Ozone

The toxicity of ozone has been summarized in a comprehensive work on the subject.[7] A deep lung irritant, ozone causes pulmonary edema, which can be fatal. It is also strongly irritating to the upper respiratory system and eyes and is largely responsible for the unpleasantness of photochemical smog. A level of 1 ppm of ozone in air has a distinct odor and inhalation of such air causes severe irritation and headache.

Like nitrogen dioxide (see Section 7.3) and ionizing radiation, ozone in the body produces free radicals that can be involved in destructive oxidation processes, such as lipid peroxidation or reaction with sulfhydryl (–SH) groups. Exposure to ozone can cause chromosomal damage. Radical-scavenging compounds, antioxidants, and compounds containing sulfhydryl groups can protect organisms from the effects of ozone.

7.5. THE HALOGENS

This section discusses the toxicities of the elemental **halogens** — fluorine, chlorine, bromine and iodine. As noted in Chapter 5, the elemental forms of the halogens are discussed here because they are chemically and toxicologically similar to many of their compounds, such as the interhalogen compounds. The toxicities of halogen compounds are discussed in the next two sections.

Chlorine

Elemental chlorine (Cl_2, mp -101°C, bp -34.5°C) is a greenish-yellow gas that is produced industrially in large quantities for numerous uses, such as the production of organochlorine solvents (see Chapter 11) and water disinfection. Liquified Cl_2 is shipped in large quantities in railway tank cars and human exposure to chlorine from transportation accidents is not uncommon.

Chlorine was the original poison gas used in World War I. It is a strong oxidant and reacts with water to produce an acidic oxidizing solution by the following reactions:

$$Cl_2 + H_2O \rightleftharpoons HCl + HOCl \quad (7.7)$$

$$Cl_2 + H_2O \rightleftharpoons 2HCl + \{O\} \quad (7.8)$$

where HOCl is oxidant hypochlorous acid and {O} stands for nascent oxygen (in a chemical sense regarded as freshly generated, highly reactive oxygen atoms). When chlorine reacts in the moist tissue lining the respiratory tract, the effect is quite damaging to the tissue. Levels of 10-20 ppm of chlorine gas in air cause immediate irritation to the respiratory tract and brief exposure to 1,000 ppm of Cl_2 can be fatal. Because of its intensely irritating properties, chlorine is not an insidious poison, and exposed individuals will rapidly seek to get away from the source if they are not immediately overcome by the gas.

Fluorine

Fluorine (F_2, mp -218°C, bp -187°C) is a pale yellow gas produced from calcium fluoride ore by first liberating hydrogen fluoride with sulfuric acid, then electrolyzing the HF in a 4:1 mixture with potassium fluoride, KF, as shown in the reaction:

$$2HF(\text{molten KF}) \xrightarrow{\text{Direct current}} H_2(\text{cathode}) + F_2(\text{anode}) \qquad (7.9)$$

Of all the elements, fluorine is the most reactive and the most electronegative (a measure of tendency to acquire electrons). In its chemically combined form it always has an oxidation number of -1. Fluorine has numerous industrial uses, such as the manufacture of UF_6, a gas used to enrich uranium in its fissionable isotope, uranium-235. Fluorine is used to manufacture uranium hexafluoride, UF_6, a dielectric material contained in some electrical and electronic apparatus. A number of organic compounds contain fluorine, particularly the chlorofluorocarbons used as refrigerants, and organofluorine polymers, such as DuPont's Teflon.

Given elemental fluorine's extreme chemical reactivity, it is not surprising that F_2 is quite toxic. It is classified as "a most toxic irritant."[4] It strongly attacks skin and the mucous membranes of the nose and eyes.

Bromine

Bromine (Br_2, mp -7.3°C, bp 58.7°C) is a dark red liquid. It is prepared commercially from elemental chlorine and bromide ion in bromide brines by the reaction:

$$Cl_2 + 2Br^- \rightarrow 2Cl^- + Br_2 \quad (7.10)$$

and the elemental bromine product is swept from the reaction mixture with steam. The major use of elemental bromine is for the production of organobromine compounds such as 1,1-dibromoethane, formerly widely used as a grain and soil fumigant for insect control and as a component of leaded gasoline for scavenging lead from engine cylinders.

Bromine is toxic when inhaled or ingested. Like chlorine and fluorine, it is an irritant to the respiratory tract and eyes because it attacks their mucous membranes. Pulmonary edema may result from severe bromine poisoning. The severely irritating nature of bromine causes a withdrawal response in its presence, thereby limiting exposure.

Iodine

Elemental iodine (I_2, solid, sublimes at 184°C) consists of violet-black rhombic crystals with a lustrous metallic appearance. More irritating to the lungs than bromine or chlorine, its general effects are similar to these elements. Exposure to iodine is limited by its low vapor pressure compared to liquid bromine or gaseous chlorine or fluorine.

7.6. HYDROGEN HALIDES

The hydrogen halides are compounds with the general formula HX,where X is F, Cl, Br, or I. They are all gases and all are relatively toxic. Because of their abundance and industrial uses, HF and HCl have the greatest toxicological significance of these gases.

Hydrogen Fluoride

Hydrogen fluoride, (HF, mp -83.1°C, bp 19.°C) may be in the form of either a clear, colorless liquid or gas. It forms corrosive fumes when exposed to the atmosphere. The major commercial application of hydrogen fluoride is as an alkylating catalyst in petroleum refining. Hydrogen fluoride in aqueous solution is called **hydrofluoric acid**, which contains 30–60% HF by mass. Hydrofluoric acid must be kept in plastic containers because it vigorously attacks glass and other materials containing silica (SiO_2), producing gaseous silicon tetrafluoride, SiF_4. Hydrofluoric acid is used to etch glass and clean stone.

Both hydrogen fluoride and hydrofluoric acid, referred to collectively as HF, are extreme irritants to any part of the body that they contact. Exposed areas heal poorly, gangrene may develop, and ulcers can occur in affected areas of the upper respiratory tract.

The toxic nature of fluoride ion, F^-, is not confined to its presence in HF. It is toxic in soluble fluoride salts, such as NaF. At relatively low levels, such as about 1 ppm used in some drinking water supplies, fluoride prevents tooth decay. At excessive levels fluoride causes **fluorosis**, a condition characterized by bone abnormalities and mottled, soft teeth. Livestock is especially susceptible to poisoning from fluoride fallout on grazing land as a result of industrial pollution. In severe cases the animals become lame and even die.

Hydrogen Chloride

Hydrogen chloride (HCl, mp -114°C, bp -84.8°C) may be encountered as a gas, pressurized liquid, or aqueous solution called **hydrochloric acid**, commonly denoted simply as HCl. This compound is colorless in the pure state and in aqueous solution. Hydrochloric acid as a saturated solution containing 36% HCl is a major industrial chemical with U.S. production of about 2.3 million tons per year. This chemical is used for chemical and food manufacture, acid treatment of oil wells to increase crude oil flow, and metal processing.

Hydrogen chloride is not nearly as toxic as HF, although inhalation can cause spasms of the larynx as well as pulmonary edema and even death at high levels. Because of its high affinity for water, HCl vapor tends to dehydrate tissue of the eyes and respiratory tract. Hydrochloric acid is a natural physiological fluid found as a dilute solution in the stomachs of humans and other animals.

Hydrogen Bromide and Hydrogen Iodide

Hydrogen bromide (HBr, mp -87°C, bp -66.5°C) and **hydrogen iodide** (HI, mp -50.8°C, bp -35.4°C) are both pale yellow or colorless gases, although contamination by their respective elements tends to impart some color to these compounds. Both are very dense gases, 3.5 g/L for HBr and 5.7 g/L for HI at 0°C and atmospheric pressure. These compounds are used much less than HCl. Both are irritants to the skin and eyes and to the oral and respiratory mucous membranes.

7.7. INTERHALOGEN COMPOUNDS AND HALOGEN OXIDES

Halogens form compounds among themselves and with oxygen. Some of these compounds are important in industry and toxicologically. Some of the more important such compounds are discussed here.

Interhalogen Compounds

Fluorine is a sufficiently strong oxidant to oxidize chlorine, bromine, and iodine, whereas chlorine can oxidize bromine and iodine. The compounds thus formed are called interhalogen compounds.[8] The major interhalogen compounds are listed in Table 7.1.

The liquid interhalogen compounds are usually described as "fuming liquids." For the most part interhalogen compounds exhibit extreme reactivity. They react with water or steam to produce hydrohalic acid solutions (HF, HCl) and nascent oxygen {O}. They tend to be potent oxidizing agents for organic matter and oxidizable inorganic compounds. These chemical properties are reflected in the toxicities of the interhalogen compounds. Too reactive to enter biological systems in their original chemical state, they tend to be powerful corrosive irritants that acidify, oxidize, and dehydrate tissue. The skin, eyes and mucous membranes of the mouth, throat, and pulmonary systems are susceptible to attack by interhalogen compounds. In some respects the toxicities of the interhalogen compounds resemble the toxic properties of the elemental forms of the elements from which they are composed. The by-products of chemical reactions of the interhalogen compounds — such as HF from fluorine compounds — pose additional toxicological hazards.

Halogen Oxides

The oxides of the halogens tend to be unstable and reactive. Although these compounds are called oxides, it is permissible to call

Table 7.1. The Major Interhalogen Compounds

Compound name and formula	Physical properties
Chlorine monofluoride, ClF	Colorless gas, mp -154°C, bp 101°C
Chlorine trifluoride, ClF_3	Colorless gas, mp -83°C, bp 12°C
Bromine monofluoride, BrF	Pale brown gas, bp 20°C
Bromine trifluoride, BrF_3	Colorless liquid, mp 8.8°C, bp 127°C
Bromine pentafluoride, BrF_3	Colorless liquid, mp -61.3°C, bp 40°C
Bromine monochloride, BrCl	Red/yellow highly unstable liquid and gas
Iodine trifluoride, IF_3	Yellow solid decomposing at 28°C
Iodine pentafluoride, IF_5	Colorless liquid, mp 9.4°C, bp 100 °C
Iodine heptafluoride, IF_7	Colorless sublimable solid, mp 5.5°C
Iodine monobromide, IBr	Gray sublimable solid, mp 42°C
Iodine monochloride, ICl	Red-brown solid alpha form, mp 27°C, bp 9°C
Iodine pentabromide, IBr_5	Crystalline solid
Iodine tribromide, IBr_3	Dark brown liquid
Iodine trichloride, ICl_3	Orange-yellow solid subliming at 64°C
Iodine pentachloride, ICl_5	—

the ones containing fluorine "fluorides" because fluorine is more electronegative than oxygen. The major halogen oxides are listed in Table 7.2.

Commercially the most important of the halogen oxides is chlorine dioxide, which offers some advantages over chlorine as a water disinfectant. It is also employed for odor control and bleaching wood pulp.[9] Because of its extreme instability, chlorine dioxide is manufactured on the site where it is used.

Table 7.2. Major Oxides of the Halogens

Compound name and formula	Physical properties
Fluorine monoxide, OF_2	Colorless gas, mp -224°C, bp -145°C
Chlorine monoxide, Cl_2O	Orange gas, mp -20°C, bp 2.2°C
Chlorine dioxide, ClO_2	Orange gas, mp -59°C, bp 9.9°C
Chlorine heptaoxide, Cl_2O_7	Colorless oil, mp -91.5°C, bp 82°C
Bromine monoxide, Br_2O	Brown solid, decomp. -18°C
Bromine dioxide, BrO_2	Yellow solid, decomp. 0°C
Iodine dioxide, IO_2	Yellow solid
Iodine pentoxide, I_2O_5	Colorless oil, decomp. 325°C

For the most part, the halogen oxides are highly reactive toxic substances. Their toxicity and hazard characteristics are similar to those of the interhalogen compounds, which were described previously in this section.

Hypochlorous Acid and Hypochlorites

The halogens form several oxyacids and their corresponding salts. Of these, the most important is hypochlorous acid (HOCl) formed by the following reaction:

$$Cl_2 + H_2O \rightleftharpoons HCl + HOCl \tag{7.11}$$

Hypochlorous acid and hypochlorites are used for bleaching and disinfection. They produce active (nascent) oxygen ({O}) as shown by the reaction below, and the resulting oxidizing action is largely responsible for the toxicity of hypochloric acid and hypochlorites as irritants to eye, skin, and mucous membrane tissue.

$$HClO \rightarrow H^+ + Cl^- + \{O\} \tag{7.12}$$

Perchlorates

Perchlorates are the most oxidized of the salts of the chloro oxyacids. Although perchlorates are not particularly toxic, ammonium perchlorate (NH_4ClO_4) should be mentioned because it is a powerful oxidizer and reactive chemical produced in large quantities as a fuel oxidizer in solid rocket fuels. Each of the U.S. space shuttle booster rockets contains about 350,000 kg of ammonium perchlorate in its propellant mixture. As of 1988, U.S. consumption of ammonium perchlorate for rocket fuel uses was of the order of 24 million kg/year. In May 1988, a series of massive explosions in Henderson, Nevada, demolished one of only two plants producing ammonium perchlorate for the U.S. space shuttle, MX missile, and other applications, so that supplies were severely curtailed.[10]

The toxicological hazard of perchlorate salts depends upon the cation in the compound. In general the salts should be considered as skin irritants and treated as such.

7.8. NITROGEN COMPOUNDS OF THE HALOGENS

Azides

The halogen azides are compounds with the general formula XN_3, where X is one of the halogens. These compounds are extremely reactive and can be spontaneously explosive. Their reactions with water can produce toxic fumes of the elemental halogen, acid (e.g., HCl), and NO_x. The compound vapors are irritants.

Nitrogen Halides

The general formula of the nitrogen halides is N_nX_x, where X is F, Cl, Br or I. A list of nitrogen halides is presented in Table 7.3.

The nitrogen halides are considered to be very toxic, largely as irritants to eyes, skin, and mucous membranes. Direct exposure to nitrogen halide compounds tends to be limited because of their reactivity, which may destroy the compound before exposure. Nitrogen triiodide is so reactive that even a "puff of air" can detonate it.[4]

Monochloramine and Dichloramine

The substitution of Cl for H on ammonia can be viewed as a means of forming nitrogen trichloride (Table 7.3), monochloramine, and dichloramine. The formation of the last two compounds from ammonium ion in water is shown by Reactions 7.13 and 7.14.

Table 7.3. Nitrogen Halides

Compound name and formula	Physical properties
Nitrogen trifluoride, NF_3	Colorless gas, mp -209°C, bp -129°C
Nitrogen trichloride, NCl_3	Volatile yellow oil, melting below -40°C, boiling below 71°C, exploding around 90°C
Nitrogen tribromide, NBr_3	Solid crystals
Nitrogen triiodide, NI_3	Black crystalline explosive substance
Tetrafluorohydrazine, N_2F_4	—

$$NH_4^+ + HOCl \rightarrow H^+ + H_2O + \underset{\text{monochloramine}}{NH_2Cl} \tag{7.13}$$

$$NH_2Cl + HOCl \rightarrow H_2O + \underset{\text{dichloramine}}{NHCl_2} \tag{7.14}$$

The chloramines are disinfectants in water and are formed deliberately in the purification of drinking water to provide **combined available chlorine**. Although combined available chlorine is a weaker disinfectant than water containing Cl_2, HOCl, and OCl^-, it is retained longer in the water system for longer-lasting disinfection.

7.9. INORGANIC COMPOUNDS OF SILICON

Because of its use in semiconductors, silicon has emerged as a key element in modern technology. Concurrent with this phenomenon has been an awareness of the toxicity of silicon compounds, many of which, fortunately, have relatively low toxicities. This section covers the toxicological aspects of inorganic silicon compounds.

Silica

The silicon compound that has probably caused the most illness in humans is **silica**, SiO_2. Silica is a hard mineral substance known as quartz in the pure form and occurring in a variety of minerals such as sand, sandstone, and diatomaceous earth. Because of silica's occurrence in a large number of common materials that are widely used in construction, sand blasting, refractories manufacture, and many other industrial applications, human exposure to silica dust is widespread. Such exposure causes a condition called **silicosis**, a type of pulmonary fibrosis, one of the most common disabling conditions that result from industrial exposure to hazardous substances. Silicosis causes fibrosis and nodules in the lung, lowering lung capacity and making the subject more liable to pulmonary diseases, such as pneumonia. Severe cases of silicosis can cause death from insufficient oxygen or from heart failure.

Asbestos

Asbestos describes a group of silicate minerals, such as those of the serpentine group, approximate formula $Mg_3P(Si_2O_5)(OH)_4$, which occur as mineral fibers. Asbestos has many properties, such as insulating abilities and heat resistance, that have given it numerous uses. It has been used in structural materials, brake linings, insulation, and pipe manufacture. Unfortunately, inhalation of asbestos damages the lungs and results in a characteristic type of lung cancer in some exposed subjects. The three major pathological conditions caused by the inhalation of asbestos are asbestosis (a pneumonia condition), mesothelioma (tumor of the mesothelial tissue lining the chest cavity adjacent to the lungs), and bronchogenic carcinoma (cancer originating with the air passages in the lungs). Because of these health effects, uses of asbestos have been severely curtailed and widespread programs have been undertaken to remove asbestos from buildings.

The toxicity of asbestos has been the subject of extensive investigation. The current status of knowledge of asbestos toxicity is summarized in a book on the subject.[11]

Silanes

Compounds of silicon with hydrogen are called **silanes**. The simplest of these is silane, SiH_4. Disilane is H_3SiSiH_3. Numerous organic silanes exist in which alkyl moieties are substituted for H.

In addition to SiH_4, the inorganic silanes produced for commercial use are dichloro- and trichlorosilane, SiH_2Cl_2 and $SiHCl_3$, respectively. These compounds are used as intermediates in the synthesis of organosilicon compounds and in the production of high-purity silicon for semiconductors. Several kinds of inorganic compounds derived from silanes have potential uses in the manufacture of photovoltaic devices for the direct conversion of solar energy to electricity. In general, not much is known about the toxicities of silanes. Silane itself burns readily in air. Chlorosilanes are irritants to eye, nasal, and lung tissue.

Silicon Halides and Halohydrides

Four **silicon tetrahalides** with the general formula SiX_4 are known to exist. Of these, only silicon tetrachloride, $SiCl_4$, is produced in significant quantities. It is used to manufacture fumed silica (finely divided SiO_2). In addition, numerous **silicon halohydrides** with the general formula $H_{4-x}SiX_x$ have been synthesized. The commercially important compound of this type is trichlorosilane, $HSiCl_3$, which is used to manufacture organotrichlorosilanes and elemental silicon for semiconductors.

Both silicon tetrachloride and trichlorosilane are fuming liquids with suffocating odors. They both react with water to give off HCl vapor.

7.10. INORGANIC PHOSPHORUS COMPOUNDS

Phosphine

Phosphine (PH_3, mp -132°C, bp -88°C) is a colorless gas that undergoes autoignition at 100°C. It is used for the synthesis of organophosphorus compounds. Its inadvertent production in chemical syntheses involving other phosphorus compounds is a potential hazard in industrial processes and in the laboratory. Phosphine gas is very toxic and can be fatal. Symptoms of exposure include fatigue, vomiting, and difficult, painful breathing. Phosphine is a pulmonary tract irritant and central nervous system depressant.

Phosphorus Pentoxide

The oxide most commonly formed by the combustion of elemental white phosphorus and many phosphorus compounds is P_4O_{10}. As an item of commerce this compound is usually called **phosphorus pentoxide**. When produced from the combustion of elemental phosphorus (see Reaction 5.7), it is a fluffy white powder that removes water from air to form syrupy orthophosphoric acid:

$$P_4O_{10} + 6H_2O \rightarrow 4H_3PO_4 \quad (7.15)$$

Because of its dehydrating action and formation of acid, phosphorus pentoxide is a corrosive irritant to skin, eyes and mucous membranes.

Phosphorus Halides

Phosphorus forms halides with the general formulas PX_3 and PX_5. Typical of such compounds are phosphorus trifluoride (PF_3), a colorless gas (mp -152°C, bp -102°C), and phosphorus pentabromide (PBr_5,) a yellow solid that decomposes at approximately 100°C. Of these compounds the most important commercially is phosphorus pentachloride, used as a catalyst in organic synthesis, as a chlorinating agent and as a raw material to make phosphorus oxychloride ($POCl_3$). Phosphorus halides react violently with water to produce the corresponding hydrogen halides and oxo phosphorus acids as shown by the following reaction of phosphorus pentachloride:

$$PCl_5 + 4H_2O \rightarrow H_3PO_4 + 5HCl \quad (7.16)$$

Largely because of their acid-forming tendencies, the phosphorus halides are strong irritants to eyes, skin, and mucous membranes, and should be regarded as very toxic.

Phosphorus Oxyhalides

Phosphorus oxyhalides with the general formula POX_3 are known for fluoride, chloride, and bromide. Of these, the one with commercial uses is phosphorus oxychloride ($POCl_3$). Its uses are similar to those of phosphorus trichloride, acting in chemical synthesis as a chlorinating agent and for the production of organic chemical intermediates. It is a faintly yellow fuming liquid, mp 1°C, bp 105°C.

It reacts with water to form hydrochloric acid and phosphonic acid (H_3PO_3). The liquid evolves toxic vapors and it is a strong irritant to the eyes, skin, and mucous membranes.

7.11. INORGANIC COMPOUNDS OF SULFUR

One of the elements essential for life, sulfur is a constituent of several of the more important toxic inorganic compounds. The common elemental form of yellow crystalline or powdered sulfur, S_8, has a low toxicity, although chronic inhalation of it can irritate mucous membranes.

Hydrogen Sulfide

Hydrogen sulfide (H_2S) is a colorless gas (mp -86°C, bp -61°C) with a foul rotten-egg odor. It is produced in large quantities as a by-product of coal coking and petroleum refining and massive quantities are removed in the cleansing of sour natural gas. It is a major source of elemental sulfur by a process that involves oxidation of part of the H_2S to SO_2 followed by the Claus reaction:

$$2H_2S(g) + SO_2(g) \rightarrow 2H_2O(l) + 3S(s) \qquad (7.17)$$

Hydrogen sulfide is a very toxic substance, which in some cases can cause a fatal response more rapidly even than hydrogen cyanide. It affects the central nervous system, causing symptoms that include headache, dizziness, and excitement. Rapid death occurs at exposures to air containing more than about 1000 ppm H_2S, and somewhat lower exposures for about 30 minutes can be lethal. Death results from asphyxiation as a consequence of respiratory system paralysis. Accidental poisonings by hydrogen sulfide are not uncommon. In the most notorious such case, 22 people were killed in 1950 in Poza Rica, Mexico, when a flare used to "dispose" of hydrogen sulfide from natural gas by burning it to sulfur dioxide became extinguished, releasing large quantities of H_2S and asphyxiating victims as they slept. In 1975 at Denver City, Texas, nine people were killed from hydrogen sulfide blown out of a secondary petroleum recovery well. There are numerous effects of chronic H_2S poisoning, including general debility.

Sulfur Dioxide and Sulfites

Sulfur dioxide (SO_2) is an intermediate in the production of sulfuric acid. It is a common air pollutant produced by the combustion of pyrite (FeS_2) in coal and organically bound sulfur in coal and fuel oil as shown by the two following reactions:

$$4FeS_2 + 11O_2 \rightarrow 2Fe_2O_3 + 8SO_2 \quad (7.18)$$

$$S(\text{organic, in fuel}) + O_2 \rightarrow SO_2 \quad (7.19)$$

These sources add millions of tons of sulfur dioxide to the global atmosphere annually and are largely responsible for acid rain. Sulfur dioxide is an irritant to the eyes, skin, mucous membranes, and respiratory system. As a water-soluble gas, it is largely removed in the upper respiratory tract (see Section 3.4).

Dissolved in water, sulfur dioxide produces **sulfurous acid**, H_2SO_3; **hydrogen sulfite ion**, HSO_3^-; and **sulfite ion**, SO_3^{2-}. Sodium sulfite (Na_2SO_3) has been used as a chemical food preservative, although some individuals are hypersensitive to it.

Sulfuric Acid

Sulfuric acid is number one in synthetic chemical production. It is used to produce phosphate fertilizer, high octane gasoline, and a wide variety of inorganic and organic chemicals. Large quantities are consumed to pickle steel (cleaning and removal of surface oxides); disposal of spent pickling liquor can be a problem.

Sulfuric acid is a severely corrosive poison and dehydrating agent in the concentrated liquid form. It readily penetrates skin to reach subcutaneous tissue and causes tissue necrosis with effects resembling those of severe thermal burns. Sulfuric acid fumes and mists can act as irritants to eye and respiratory tract tissue. Industrial exposure has caused tooth erosion in workers.

Miscellaneous Inorganic Sulfur Compounds

A large number of inorganic sulfur compounds including halides and salts are widely used in industry. The more important of these are listed in Table 7.4.

Table 7.4. Inorganic Sulfur Compounds

Compound name and formula	Properties
Sulfur	
Monofluoride, S_2F_2	Colorless gas, mp -104°C, bp -99°C, toxicity similar to HF
Tetrafluoride, SF_4	Gas, bp -40°C, mp -124°C, powerful irritant
Hexafluoride, SF_6	Colorless gas, mp -51°C, surprisingly nontoxic when pure, but often contaminated with toxic lower fluorides
Monochloride, S_2Cl_2	Oily, fuming orange liquid, mp -80°C, bp 138°C, strong irritant to eyes, skin, and lungs
Tetrachloride, SCl_4	Brownish/yellow liquid/gas, mp -30°C, Decom. below 0°C, irritant
Trioxide, SO_3	Solid anhydride of sulfuric acid (see toxic effects above), reacts with moisture or steam to produce sulfuric acid
Sulfuryl chloride, SO_2Cl_2	Colorless liquid, mp -54°C, bp 69°C, used for organic synthesis, corrosive toxic irritant
Thionyl chloride, $SOCl_2$	Colorless-to-orange fuming liquid, mp -105°C, bp 79°C, toxic corrosive irritant
Carbon oxysulfide, COS[1]	Volatile liquid byproduct of natural gas or petroleum refining, toxic narcotic
Carbon disulfide, CS_2 [1]	Colorless liquid, industrial chemical, narcotic and CNS anesthetic

[1] Carbon oxysulfide and carbon disulfide may also be classed as organic compounds, and their toxicological properties are discussed with organosulfur compounds in Chapter 12.

LITERATURE CITED

1. Seiler, Hans G., and Helmut Sigel, Eds., *Handbook on Toxicity of Inorganic Compounds*, Marcel Dekker, New York, 1987.

2. Gosselin, Robert E., Roger P. Smith, and Harold C. Hodge, "Cyanide," in *Clinical Toxicology of Commercial Products*, 5th ed., Williams and Wilkins, Baltimore/London, 1984, pp. III-123–III-130.

3. Hodgson, Ernest, and Patricia E. Levi, *Modern Toxicology*, Elsevier, New York, 1987.

4. Sax, N. Irving, *Dangerous Properties of Industrial Materials*, 5th ed., Van Nostrand Reinhold, New York 1979.

5. Finlayson-Pitts, Barbara J., and James N. Pitts, Jr., *Atmospheric Chemistry*, John Wiley and Sons, New York, 1986.

6. Timbrell, John A., *Principles of Biochemical Toxicology*, Taylor and Francis, Ltd., London, 1982.

7. Lee, S.D., M. G. Mustafa, and M. A. Mehlman, Eds., *International Symposium on the Biochemical Effects of Ozone and Related Photochemical Oxidants*, Vol. V in *Advances in Modern Environmental Toxicology*, Princeton University Press, Princeton, NJ, 1983.

8. Gillespie, Ronald J., David A. Humphries, N. Colin Baird, and Edward A. Robinson, *Chemistry*, Allyn and Bacon, Boston, 1986.

9. Noack, Manfred G., and Richard L. Doerr, "Chlorous Acid, Chlorites, and Chlorine Dioxide," in *Kirk-Othmer Concise Encyclopedia of Chemical Technology*, Wiley-Interscience, New York, 1985.

10. Seltzer, Richard, "Shortage of Rocket Fuel Oxidizer Possible," *Chemical and Engineering News*, May 26, 1988, p. 5.

11. Fisher, Gerald L., and Michael A. Gallo, Eds., *Asbestos Toxicity*, Marcel Dekker, New York, 1988.

8

Toxic Organic Compounds and Hydrocarbons

8.1. INTRODUCTION

The fundamentals of organic chemistry are reviewed in Chapter 2. The present chapter is the first of seven that discuss the toxicological chemistry of organic compounds that are largely of synthetic origin. Since the vast majority of the several million known chemical compounds are organic — most of them toxic to a greater or lesser degree — the toxicological chemistry of organic compounds covers an enormous area. Specifically, this chapter discusses hydrocarbons, which are organic compounds composed only of carbon and hydrogen and are in a sense the simplest of the organic compounds. Hydrocarbons occur naturally in petroleum, natural gas, and tar sands and they can be produced by pyrolysis of coal and oil shale or by chemical synthesis from H_2 and CO.

8.2. CLASSIFICATION OF HYDROCARBONS

For purposes of discussion of hydrocarbon toxicities in this chapter, hydrocarbons will be grouped into the five categories of (1) **alkanes**, (2) **unsaturated nonaromatic** hydrocarbons, (3) **aromatic** hydrocarbons (understood to have only one or two linked aromatic rings in their structures), (4) **polycyclic** aromatic hydrocarbons with multiple rings, and (5) **mixed** hydrocarbons containing combinations of two or more of the preceding types. These classifications are summarized with examples in Figure 8.1.

Alkanes

$$\begin{array}{c} H \\ | \\ H-C-H \\ | \\ H \end{array} \qquad \begin{array}{cccc} H & CH_3 & CH_3 & H \\ | & | & | & | \\ H-C- & C- & C- & C-H \\ | & | & | & | \\ H & CH_3 & H & H \end{array}$$

Methane 2,2,3,-Trimethylbutane

Unsaturated nonaromatic

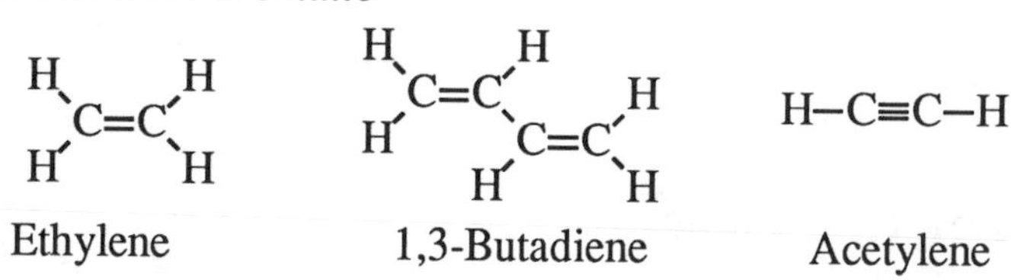

One/two-ring aromatic

Polycyclic aromatic

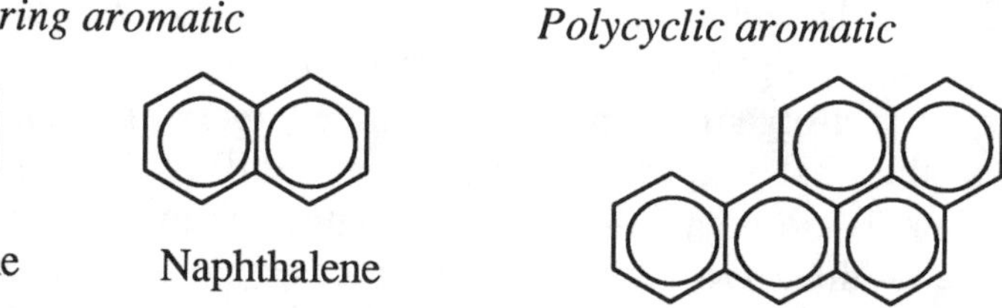

Benzo(a)pyrene

Mixed hydrocarbons

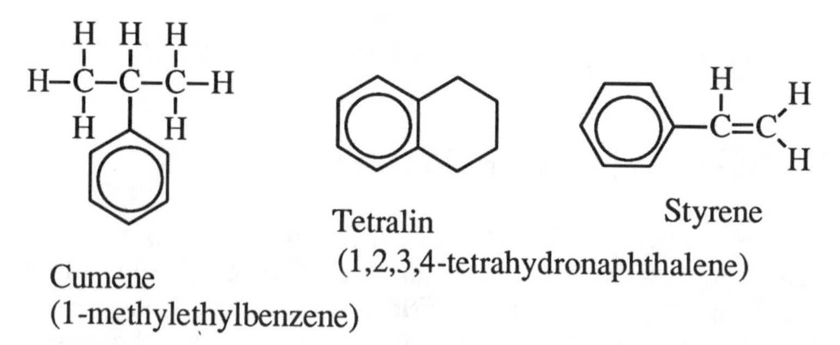

Figure 8.1. Hydrocarbons classified for discussion of their toxicological chemistry.

Alkanes

Alkanes, also called **paraffins** or **aliphatic hydrocarbons**, are hydrocarbons in which the C atoms are joined by single covalent bonds (sigma bonds) consisting of two shared electrons (see Section 2.4). As shown by the examples in Figure 8.1 and Chapter 2, Figure 2.5, alkanes may exist as straight chains or branched chains. They may also exist as cyclic structures, for example, as in cyclohexane (C_6H_{12}). Each cyclohexane molecule consists of 6 carbon atoms (each with 2 H atoms attached) in a ring. The general molecular formula for straight- and branched-chain alkanes is C_nH_{2n+2}, and that of a cyclic alkane is C_nH_{2n}. The names of alkanes having from 1 to 10 carbon atoms per molecule are (1) methane, (2) ethane, (3) propane, (4) butane, (5) pentane, (6) hexane, (7) heptane, (8) octane, (9) nonane, and (10) decane. These names may be prefixed by "*n*–" to denote a straight-chain alkane. The same base names are used to designate substituent groups on molecules; for example, a straight-chain 4-carbon alkane group (derived from butane) attached by an end carbon to a molecule is designated as an *n*-butyl group.

Alkanes undergo a number of chemical reactions, two classes of which should be mentioned here. The first of these is **oxidation** with molecular oxygen in air as shown for the following combustion reaction of propane:

$$C_3H_8 + 5O_2 \rightarrow 3CO_2 + 4H_2O + \text{heat} \qquad (8.1)$$

Such reactions can pose flammability and explosion hazards. Another hazard occurs during combustion in an oxygen-deficient atmosphere or in an automobile engine, in which significant quantities of toxic carbon monoxide (CO) are produced.

The second major type of alkane reaction that should be considered here consists of **substitution reactions** in which one or more H atoms on an alkane are replaced by atoms of another element. Most commonly the H is replaced by a halogen, usually chlorine, to yield **organohalide** compounds; when chlorine is the substituent the product is called an **organochlorine** compound. An example of this kind of reaction is that of methane with chlorine to give carbon tetrachloride, Reaction 8.2. Organohalide compounds are of great toxicological significance and are discussed in Chapter 11.

$$\text{H-}\underset{\text{H}}{\overset{\text{H}}{\text{C}}}\text{-H} + 4Cl_2 + \rightarrow \text{Cl-}\underset{\text{Cl}}{\overset{\text{Cl}}{\text{C}}}\text{-Cl} + 4HCl \quad (8.2)$$

Unsaturated Nonaromatic Hydrocarbons

Unsaturated hydrocarbons are those that have multiple bonds, each involving more than 2 shared electrons, between carbon atoms. Such compounds are usually **alkenes** or **olefins** that have double bonds consisting of 4 shared electrons as shown for ethylene and 1,3-butadiene in Figure 8.1. Triple bonds consisting of 6 shared electrons are also possible as illustrated by acetylene in the same figure.

Alkenes may undergo **addition reactions** in which pairs of atoms are added across unsaturated bonds as shown in the following reaction of ethylene with hydrogen to give ethane:

$$H_2C{=}CH_2 + \text{H-H} \rightarrow \text{H-}\underset{\text{H}}{\overset{\text{H}}{\text{C}}}\text{-}\underset{\text{H}}{\overset{\text{H}}{\text{C}}}\text{-H} \quad (8.3)$$

This kind of reaction, which is not possible with alkanes, adds to the chemical and metabolic, as well as toxicological, versatility of compounds containing unsaturated bonds.

Another example of an addition reaction is that of a molecule of HCl gas to one of acetylene to yield vinyl chloride:

$$\text{H-C}{\equiv}\text{C-H} + \text{H-Cl} \rightarrow H_2C{=}CHCl \quad (8.4)$$

The vinyl chloride product is the monomer used to manufacture polyvinylchloride plastic and is a carcinogen known to cause a rare form of liver cancer among exposed workers.

As discussed in Chapter 2, Section 2.6, compounds with double bonds can exist as geometrical isomers exemplified by the two isomers of 1,2-dichloroethylene in Figure 8.2. Although both of these compounds have the molecular formula $C_2H_2Cl_2$, the orientations of their H and Cl atoms relative to each other are different and their properties, such as melting and boiling points, are not the same. Their toxicities are both relatively low, but significantly different. The *cis*-isomer is an irritant and narcotic known to damage the liver and kidneys of experimental animals.[1] The *trans*- isomer causes weakness, tremor and cramps due to its effects on the central nervous system, as well as nausea and vomiting resulting from adverse effects on the gastrointestinal tract.

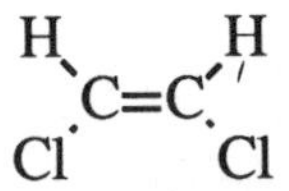

H ╲ C=C ╱ H Cl ╱ ╲ Cl	H ╲ C=C ╱ Cl Cl ╱ ╲ H
Cis –1,2–dichloroethylene, mp -80.5°C, bp 59°C	*Trans* –1,2–dichloroethylene, mp -50°C, bp 48°C

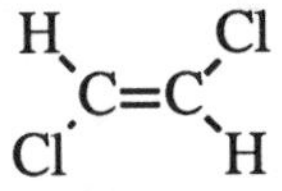

Figure 8.2. The two geometrical isomers of 1,2–dichloroethane.

Aromatic Hydrocarbons

Aromatic compounds were discussed briefly in Chapter 2, Section 2.6. The characteristics of **aromaticity** of organic compounds are numerous and are discussed at length in works on organic chemistry. These include a low hydrogen:carbon atomic ratio; C–C bonds that are quite strong and of intermediate length between such bonds in alkanes and those in alkenes; tendency to undergo substitution reactions (see Reaction 8.2) rather than the addition reactions characteristic of alkenes; and delocalization of π electrons over several carbon atoms resulting in resonance stabilization of the molecule. For more detailed explanations of these concepts the reader is referred to standard textbooks on organic chemistry. For our purposes most of the aromatic compounds discussed are those that contain single benzene rings or fused benzene rings, such as those in naphthalene or benzo(a)pyrene, shown in Figure 8.1.

An example reaction of aromatic compounds with considerable environmental and toxicological significance is the chlorination of biphenyl. Biphenyl gets its name from the fact that it consists of two **phenyl** groups (where a phenyl group is a benzene molecule less a hydrogen atom) joined by a single covalent bond. In the presence of an iron(II) chloride catalyst this compound reacts with chlorine to form a number of different molecules of polychlorinated biphenyls (PCBs) as shown in Figure 8.3. These environmentally persistent compounds are discussed in Chapter 11.

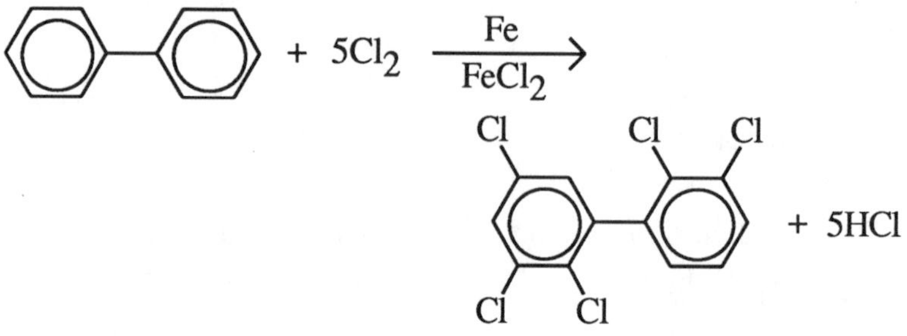

$$C_{12}H_{10} + 5Cl_2 \xrightarrow[FeCl_2]{Fe} C_{12}H_5Cl_5 + 5HCl$$

Figure 8.3. An example of a substitution reaction of an aromatic hydrocarbon compound (biphenyl) to produce an organochlorine product (2,3,5,2',3',-pentachlorobiphenyl, a PCB compound). The product is one of 210 possible congeners of PCBs, widespread and persistent pollutants found in the fat tissue of most humans and of considerable environmental and toxicological concern.

8.3. TOXICOLOGY OF ALKANES

Worker exposure to alkanes, especially the lower-molecular-mass compounds, is most likely to come from inhalation. In an effort to set reasonable values for the exposure by inhalation of vapors of solvents, hydrocarbons, and other volatile organic liquids, the American Conference of Governmental Industrial Hygienists sets **threshold limit values** (TLVs) for airborne toxicants.[2,3] The **time-weighted average exposure** (E) is calculated by the formula:

$$E = \frac{C_aT_a + C_bT_b + \cdots + C_nT_n}{8} \tag{8.5}$$

where C is the concentration of the substance in the air for a particular time T (hours), such as a level of 3.1 parts per million by volume for 1.25 hours. The 8 in the denominator is for an 8-hour day. In addition to exposures calculated by this equation, there are maximum exposures that should not be exceeded at any time, and there may be a relatively high exposure level that may be approached, but not exceeded, for brief periods of time, such as 10 minutes once each day.

"Safe" levels of air contaminants are difficult to set based upon systemic toxicologic effects. Therefore, TLV values often reflect non-systemic effects of odor, narcosis, eye irritation, and skin irritation. Because of this, comparison of TLV values is often not useful in comparing systemic toxicological effects of chemicals in the workplace.

Methane and Ethane

Methane and ethane are **simple asphyxiants**, which means that air containing high levels of these gases does not contain sufficient oxygen to support respiration. Table 8.1 shows the levels of asphyxiants in air at which various effects are observed in humans. Simple asphyxiant gases are not known to have major systemic toxicological effects, although subtle effects that are hard to detect should be considered as possibilities.

Table 8.1. Effects of Simple Asphyxiants in Air

Percent asphyxiant*	Percent oxygen, O_2*	Effect on humans
0 – 33	21 – 14	No major adverse symptoms
33 – 50	14 – 10.5	Discernible effects beginning with air hunger and progressing to impaired mental alertness and muscular coordination
50 – 75	10.5 – 5.3	Fatigue, depression of all sensations, faulty judgment, emotional instability; in later phases nausea, vomiting, prostration, unconsciousness, convulsions, coma, death
75 – 100	5.3 – 0	Death within a few minutes

* Percent by volume on a "dry" (water-vapor-free) basis

Propane and Butane

Propane has the formula C_3H_8 and butane C_4H_8. There are two isomers of butane — *n*-butane and isobutane (2-methylpropane). Propane and the butane isomers are gases at room temperature and atmospheric pressure; like methane and ethane, all three are asphyxiants. A high concentration of propane affects the central nervous system. There are essentially no known systemic toxicological effects of the two butane isomers; behavior similar to that of propane might be expected.

Pentane through Octane

The alkanes with 5 through 8 carbon atoms consist of *n*-alkanes, and there is an increasing number of branched-chain isomers with higher numbers of C atoms per molecule. For example, there are 9 isomers of heptane C_7H_{16}. These compounds are all volatile liquids under ambient conditions; the boiling points for the straight-chain isomers range from 36.1°C for *n*-pentane to 125.8°C for *n*-octane. In addition to their uses in fuels, such as in gasoline, these compounds are employed as solvents in formulations for a number of commercial products, including varnishes, glues, and inks. They are also used for the extraction of fats.

Once regarded as toxicologically almost harmless, the C_5–C_8 aliphatic hydrocarbons are now recognized as having some significant toxic effects. Exposure to the C_5–C_8 hydrocarbons is primarily via the pulmonary route, and high levels in air have killed experimental animals. Humans inhaling high levels of these hydrocarbons have become dizzy and have lost coordination as a result of central nervous system depression.

Of the C_5–C_8 alkanes, the one most commonly used for nonfuel purposes is *n*-hexane. It acts as a solvent for the extraction of oils from seeds, such as cotton seed and sunflower seed. This alkane serves as a solvent medium for several important polymerization processes and in mixtures with more polar solvents, such as furfural:

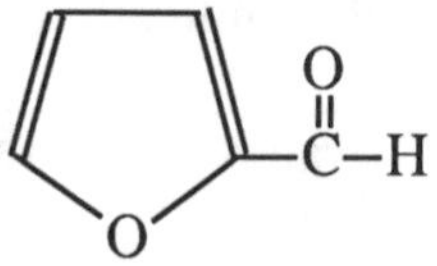

Furfural

for the separation of fatty acids.[4] **Polyneuropathy** (multiple disorders of the nervous system) has been reported in several cases of human exposure to *n*-hexane,[5] such as Japanese workers involved in the home production of sandals using a glue with *n*-hexane solvent. The workers suffered from muscle weakness and impaired sensory function of the hands and feet. Biopsy examination of nerves in leg muscles of the exposed workers showed loss of myelin (a fatty substance constituting a sheath around certain nerve fibers) and degeneration of axons (part of a nerve cell through which nerve impulses are transferred out of the cell). The symptoms of polyneuropathy were reversible, with recovery taking several years after exposure was ended.

Exposure of the skin to C_5–C_8 liquids causes dermatitis. This is the most common toxicological occupational problem associated with the use of hydrocarbon liquids in the workplace and is a consequence of the dissolution of the fat portions of the skin. In addition to becoming inflamed, the skin becomes dry and scaly.

Alkanes above Octane

Alkanes higher than C_8 are contained in kerosene, jet fuel, diesel fuel, mineral oil, and fuel oil. Kerosene, also called fuel oil No. 1, is a mixture of primarily C_8–C_{16} hydrocarbons, predominantly alkanes with a boiling point range of approximately 175–325°C. Diesel fuel is called fuel oil No. 2. The heavier fuel oils No. 3–6 are characterized by increasing viscosity, darker color and higher boiling temperatures with increasing fuel oil number. Mineral oil is a carefully selected fraction of petroleum hydrocarbons with density ranges of 0.83–0.86 g/mL for light mineral oil and 0.875–0.905 g/mL for heavy mineral oil.

The higher alkanes are not regarded as very toxic, although there are some reservations about their toxicities. Inhalation is the most common route of occupational exposure and can result in dizziness, headache, and stupor. In cases of extreme exposure, coma and death have occurred. Inhalation of mists or aspiration of vomitus containing higher alkane liquids has caused a condition known as aspiration pneumonia. Mineral oils have been reported to be carcinogenic to the skin and scrotum.[6]

Solid and Semisolid Alkanes

Semisolid petroleum jelly is a highly refined product commonly known as "vaseline," a mixture of predominantly C_{16}–C_{19} alkanes. Carefully controlled refining processes are used to remove nitrogen and sulfur compounds, resins, and unsaturated hydrocarbons. Paraffin wax is a similar product behaving as a solid. Neither petroleum jelly nor paraffin is digested or absorbed by the body.

Cyclohexane

Cyclohexane, the 6-carbon ring hydrocarbon with the molecular formula C_6H_{12}, is the most significant of the cyclic alkanes. Under ambient conditions it is a clear, volatile, highly flammable liquid. It is manufactured by the hydrogenation of benzene and is used primarily as a raw material for the synthesis of cyclohexanol and cyclohexanone through a liquid-phase oxidation with air in the presence of a dissolved cobalt catalyst.

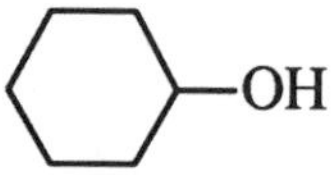

—OH Cyclohexanol   =O Cyclohexanone

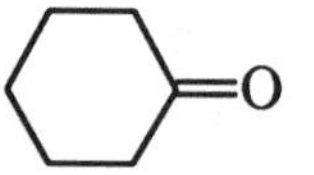

Like *n*-hexane, cyclohexane has a toxicity rating of 3, moderately toxic (see Table 1.1 for toxicity ratings). Cyclohexane acts as a weak anesthetic similar to, but more potent than, *n*-hexane. Systemic effects have not been shown in humans.

8.4. TOXICOLOGY OF UNSATURATED NONAROMATIC HYDROCARBONS

Ethylene (structure in Figure 8.1) is the most widely used organic chemical.[7] Almost all of it is consumed as a chemical feedstock for the manufacture of other organic chemicals. Polymerization of ethylene to produce polyethylene is illustrated in Figure 8.4. In addition to polyethylene, other polymeric plastics, elastomers, fibers, and resins are manufactured with ethylene as one of the ingredients. Ethylene is also the raw material for the manufacture of ethylene glycol antifreeze, solvents, plasticizers, surfactants, and coatings.

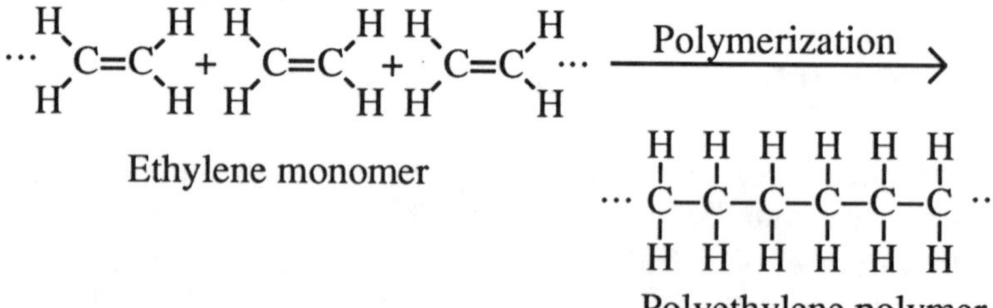

```
H       H H       H H       H
 \     /   \     /   \     /
··· C=C   +   C=C   +   C=C ···      Polymerization
 /     \   /     \   /     \        ---------------->
H       H H       H H       H

      Ethylene monomer              H  H  H  H  H  H
                                    |  |  |  |  |  |
                                ··· C—C—C—C—C—C ···
                                    |  |  |  |  |  |
                                    H  H  H  H  H  H
                                  Polyethylene polymer
```

Figure 8.4. Polymerization of ethylene to produce polyethylene.

The boiling point of ethylene is -105°C and under ambient conditions it is a colorless gas. It has a somewhat sweet odor, is highly flammable and forms explosive mixtures with air. Because of its double bond (unsaturation), ethylene is much more active than the alkanes. It undergoes addition reactions as shown in the following examples to form a number of important products:

$$\mathrm{H_2C{=}CH_2} + \mathrm{O_2} \xrightarrow{\text{Catalyst}} \underset{\text{Ethylene oxide}}{\mathrm{H_2C\!-\!CH_2}\ (\text{bridged by O})} \xrightarrow{\text{Hydrolysis}} \underset{\text{Ethylene glycol}}{\begin{array}{ccc} & \mathrm{H} & \mathrm{H} & \\ \mathrm{H-} & \mathrm{C} & \mathrm{-C} & \mathrm{-H} \\ & \mathrm{OH} & \mathrm{OH} & \end{array}} \tag{8.6}$$

$$\mathrm{H_2C{=}CH_2} + \mathrm{Br_2} \rightarrow \underset{\substack{\text{1,2-Dibromoethane} \\ \text{(ethylene dibromide)}}}{\begin{array}{cccc} & \mathrm{H} & \mathrm{H} & \\ \mathrm{Br-} & \mathrm{C} & \mathrm{-C} & \mathrm{-Br} \\ & \mathrm{H} & \mathrm{H} & \end{array}} \tag{8.7}$$

$$\mathrm{H_2C{=}CH_2} + \mathrm{Cl_2} \rightarrow \underset{\substack{\text{1,2–Dichloroethane} \\ \text{(ethylene dichloride)}}}{\begin{array}{cccc} & \mathrm{H} & \mathrm{H} & \\ \mathrm{Cl-} & \mathrm{C} & \mathrm{-C} & \mathrm{-Cl} \\ & \mathrm{H} & \mathrm{H} & \end{array}} \tag{8.8}$$

$$\mathrm{H_2C{=}CH_2} + \mathrm{HCl} \rightarrow \underset{\substack{\text{Chloroethane} \\ \text{(ethyl chloride)}}}{\begin{array}{cccc} & \mathrm{H} & \mathrm{H} & \\ \mathrm{H-} & \mathrm{C} & \mathrm{-C} & \mathrm{-Cl} \\ & \mathrm{H} & \mathrm{H} & \end{array}} \tag{8.9}$$

The products of the addition reactions shown above are all commercially, toxicologically, and environmentally important. Ethylene oxide is a highly reactive colorless gas used as a sterilizing agent, fumigant, and intermediate in the manufacture of ethylene glycol and surfactants. It is an irritant to eyes and pulmonary tract mucous membrane tissue; inhalation of it can cause pulmonary edema. Ethylene glycol is a colorless, somewhat viscous liquid used in mixtures with water as a high-boiling, low-freezing-temperature liquid (antifreeze and antiboil) in cooling systems. Ingestion of this compound causes central nervous system effects characterized by initial stimulation followed by depression. Higher doses can cause fatal kidney failure due to metabolic oxidation of ethylene glycol to oxalic acid followed by the formation of insoluble calcium oxalate which clogs the kidneys, as discussed in Section 9.2.

Ethylene dibromide has been used as an insecticidal fumigant and additive to scavenge lead from leaded gasoline combustion. During the early 1980s there was considerable concern about residues of this compound in food products and it was suspected of being a carcinogen, mutagen, and teratogen. Ethylene dichloride (bp 83.5°C) is a colorless, volatile liquid with a pleasant odor used as a soil and foodstuff fumigant. It has a number of toxicological effects, including adverse effects on the eye, liver, and kidneys and a narcotic effect on the central nervous system. Ethyl chloride seems to have similar, but much less severe, toxic effects.

Ethylene, itself, is not very toxic to animals but it is a simple asphyxiant (see Section 8.3 and Table 8.1). At high concentrations it acts as an anesthetic to induce unconsciousness. A highly flammable compound, ethylene forms dangerously explosive mixtures with air. It is phytotoxic (toxic to plants).

Propylene

Propylene (C_3H_6) is a gas with chemical, physical and toxicological properties very similar to those of ethylene. It, too, is a simple asphyxiant. Its major use is in the manufacture of polypropylene polymer, a hard, strong plastic from which are made injection-molded bottles as well as pipes, valves, battery cases, automobile body parts, and rot-resistant indoor-outdoor carpet.

1,3-Butadiene

The dialkene 1,3-butadiene is widely used in the manufacture of polymers, particularly synthetic rubber. The first synthetic rubber to be manufactured on a large scale and used as a substitute for unavailable natural rubber during World War II was a styrene-butadiene polymer.

$$\underset{\text{Styrene}}{C_6H_5-CH=CH_2} + \underset{\text{Butadiene}}{CH_2=CH-CH=CH_2} \xrightarrow{\text{Polymerization}} \underset{\text{Buna–S synthetic rubber}}{\cdots CH(C_6H_5)-CH_2-CH_2-CH=CH-CH_2\cdots} \quad (8.10)$$

Butadiene is a colorless gas under ambient conditions with a mild, somewhat aromatic odor. At lower levels the vapor is an irritant to eyes and respiratory system mucous membranes and at higher levels it can cause unconsciousness and even death. The compound boils at –4.5°C, and is readily stored and handled as a liquid. Release of the liquid can cause frostbite-like burns on exposed flesh.

Butylenes

There are four monoalkenes with the formula C_4H_8 (butylenes) as shown in Figure 8.5. All gases under ambient conditions, these compounds have boiling points ranging from -6.9°C for isobutylene to +3.8°C for *cis*-2-butene. The butylenes readily undergo isomerization (change to other isomers). They participate in addition reactions and form polymers. Their major hazard is extreme flammability. Though not regarded as particularly toxic, they are asphyxiants and have a narcotic effect when inhaled.

Alpha-Olefins

Alpha-olefins are linear alkenes with double bonds between carbons 1 and 2 in the general range of carbon chain length C_6 through about C_{18}. They are used for numerous purposes. The C_6–C_8 compounds are used as comonomers to manufacture modified polyethylene polymer and the C_{12}–C_{18} alpha-olefins are used as raw materials in the manufacture of detergents. The compounds are also used to manufacture lubricants and plasticizers. In 1986 worldwide consumption of the alpha-olefins was estimated at 800,000 metric tons estimated to rise to 1.7 million metric tons by the year 2000.[8] With such large quantities involved, due consideration needs to be given to the toxicological and occupational health aspects of these compounds.

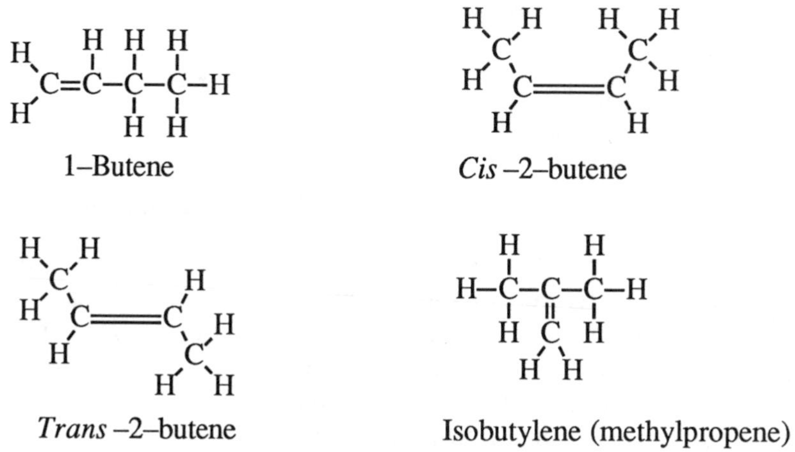

1–Butene

Cis –2–butene

Trans –2–butene

Isobutylene (methylpropene)

Figure 8.5. The four butylene compounds, formula C_4H_8.

Cyclopentadiene and Dicyclopentadiene

The cyclic dialkene cyclopentadiene has the structural formula shown below:

Cyclopentadiene

Two molecules of cyclopentadiene readily and spontaneously join together to produce dicyclopentadiene, widely used to produce polymeric elastomers, polyhalogenated flame retardants and polychlorinated pesticides. Dicyclopentadiene (mp 32.9°C, bp 166.6°C) exists as colorless crystals. It is an irritant and has narcotic effects. It is considered to have a high oral toxicity and to be moderately toxic through dermal absorption.

Acetylene

Acetylene (Figure 8.1) is widely used as a chemical raw material and fuel for oxyacetylene torches. It was once the principal raw material for the manufacture of vinyl chloride (see Reaction 8.4), but other synthetic routes are now used. Acetylene is a colorless gas with an odor resembling garlic. Though not notably toxic, it acts as an asphyxiant and narcotic and has been used for anesthesia. Exposure can cause headache, dizziness, and gastric disturbances. Some adverse effects from exposure to acetylene may be due to the presence of impurities in the commercial product.

8.5. BENZENE AND ITS DERIVATIVES

Figure 8.6 shows the structural formulas of benzene and its major hydrocarbon derivatives. These compounds are very significant in chemical synthesis, as solvents, and in unleaded gasoline formulations.

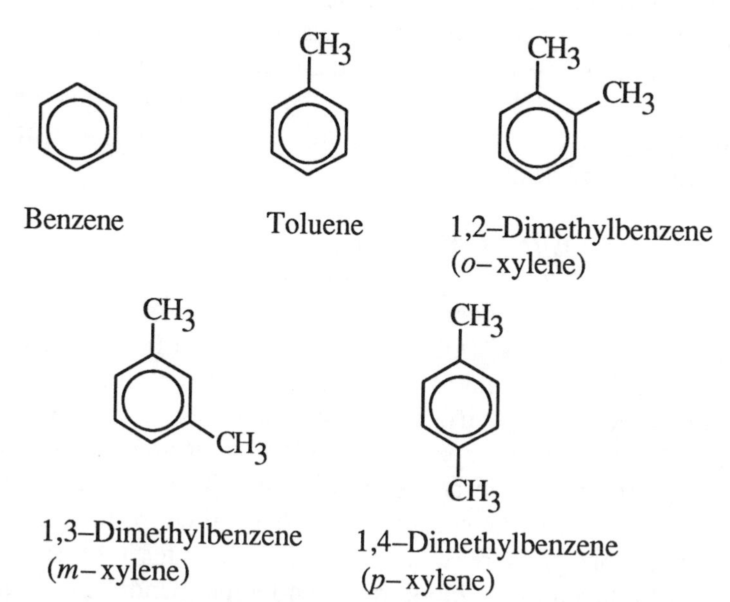

Figure 8.6. Benzene and its most common methyl-substituted hydrocarbon derivatives.

Benzene

Benzene (C_6H_6) is chemically the single most significant hydrocarbon. It is used as a starting material for the manufacture of numerous products including phenolic and polyester resins, polystyrene plastics and elastomers (through intermediate styrene, Figure 8.1), alkylbenzene surfactants, chlorobenzene compounds, insecticides, and dyes. Benzene (bp 80.1°C) is a volatile, colorless, highly flammable liquid with a characteristic odor.

Acute Toxic Effects of Benzene

Benzene has been in commercial use for over a century, and toxic effects of it have been suspected since about 1900.[9] Benzene has both acute and chronic toxicological effects.[10] It is usually absorbed as a vapor through the respiratory tract, although absorption of liquid through the skin and intake through the gastrointestinal tract are also possible. Benzene is a skin irritant, and progressively higher local exposures can cause skin redness (erythema), burning sensations, fluid accumulation (edema) and blistering. Inhalation of air containing about 64 g/m^3 of benzene can be fatal within a few minutes; about 1/10 that level of benzene causes acute poisoning within an hour, including a narcotic effect upon the central nervous system manifested progressively by excitation, depression, respiratory system failure and death.

Chronic Toxic Effects of Benzene

Of greater overall concern than the acute effects of benzene exposure are chronic effects, which are still subject to intense study. As with many other toxicants, subjects suffering from chronic benzene exposure suffer non-specific symptoms, including fatigue, headache, and appetite loss. More specifically, blood abnormalities appear in people suffering chronic benzene poisoning. The most common of these is a lowered white cell count. More detailed examination may show an abnormal increase in blood lymphocytes (colorless corpuscles introduced to the blood from the lymph glands), anemia, and decrease in the number of blood platelets required for clotting (thrombocytopenia). Some of the observed blood abnormalities may result from damage by benzene to blood marrow. Because of concerns that long-term exposure to benzene may cause preleukemia, leukemia, or cancer, the allowable levels of benzene in the workplace have been greatly reduced and substitutes such as toluene and xylene are used wherever possible.

Metabolism of Benzene

For a hydrocarbon, the water solubility of benzene is a moderately high 1.80 g/L at 25°C. The vapor is readily absorbed by blood, from which it is strongly taken up by fatty tissues. For non-metabolized benzene, the process is reversible and benzene is excreted through the lungs. Benzene metabolism occurs in the liver where it undergoes a Phase I oxidation reaction to phenol[11] as shown in Figure 8.7.

All monocyclic aromatic hydrocarbons with 6-membered rings other than benzene have substituent groups (such as the methyl group in toluene) upon which Phase I reactions may be initiated. However, the oxidative metabolism of benzene requires attachment of oxygen to the aromatic ring as shown in Figure 8.7. This is probably responsible for the unique toxicity of benzene, especially in respect to bone marrow damage. The epoxide likely reacts with cell nucleophiles, damaging or destroying the cells.

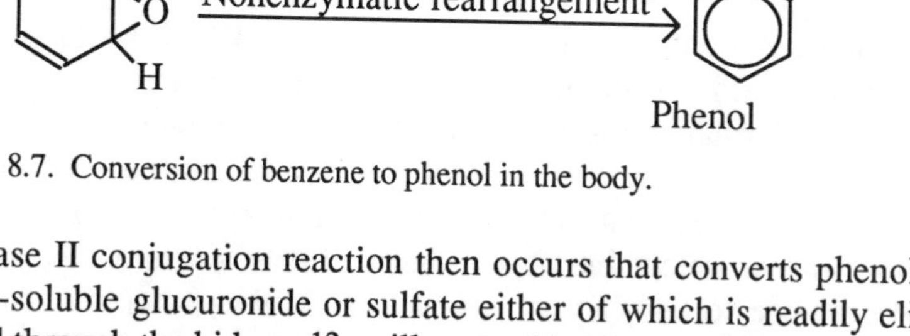

+ {O} Enzymatic epoxidation →

H O H

Benzene epoxide

H O H

Nonenzymatic rearrangement →

OH

Phenol

Figure 8.7. Conversion of benzene to phenol in the body.

A Phase II conjugation reaction then occurs that converts phenol to water-soluble glucuronide or sulfate either of which is readily eliminated through the kidneys[12] as illustrated in Figures 8.8 and 8.9.

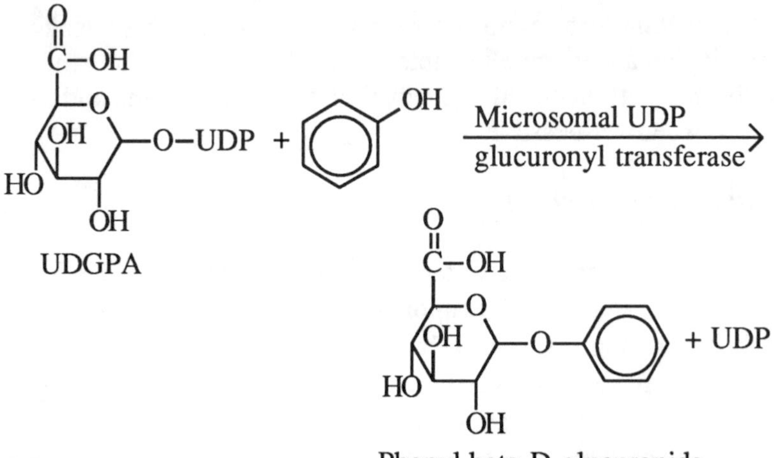

UDGPA

Microsomal UDP glucuronyl transferase

+ UDP

Phenyl beta-D-glucuronide

Figure 8.8. Formation of a glucuronide of phenol. The abbreviation UDPGA stands for uridine diphosphate glucuronic acid and UDP for uridine diphosphate.

Remainder of PAPS molecule

PAPS

+ ADP

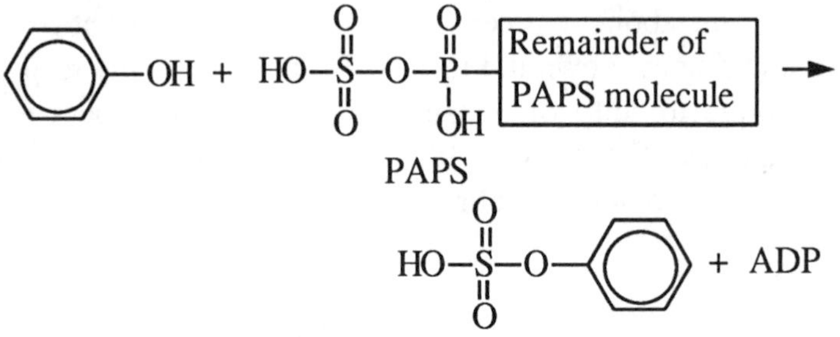

Figure 8.9. Addition reaction for the formation of excretable phenyl sulfate. The abbreviation PAPS is for adenosine 3'-phosphate-5'-phosphosulfate, a somewhat complex structure. ADP is adenosine diphosphate.

Toluene

Toluene is a colorless liquid boiling at 101.4°C. It is classified as moderately toxic through inhalation or ingestion and has a low toxicity by dermal exposure. Concentrations in ambient air up to 200 ppm usually do not result in significant symptoms, but exposure to 500 ppm may cause headache, nausea, lassitude, and impaired coordination without detectable physiological effects. At massive exposure levels, toluene acts as a narcotic, which can lead to coma.

Unlike benzene, toluene possesses an aliphatic side-chain that can be oxidized enzymatically, leading to products that are readily excreted from the body. The metabolism of toluene is thought to proceed via oxidation of the methyl group and formation of the conjugate compound hippuric acid as shown in Figure 8.10.

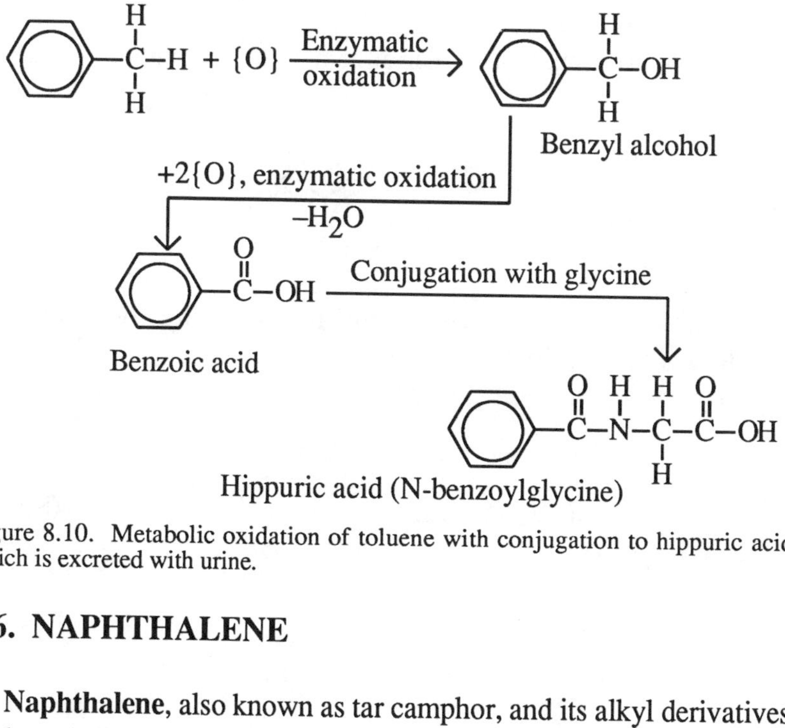

$$C_6H_5{-}CH_3 + \{O\} \xrightarrow[\text{oxidation}]{\text{Enzymatic}} C_6H_5{-}CH_2{-}OH \text{ (Benzyl alcohol)}$$

$$C_6H_5{-}CH_2{-}OH \xrightarrow[-H_2O]{+2\{O\},\ \text{enzymatic oxidation}} C_6H_5{-}C(=O){-}OH \text{ (Benzoic acid)}$$

$$C_6H_5{-}C(=O){-}OH \xrightarrow{\text{Conjugation with glycine}} C_6H_5{-}C(=O){-}NH{-}CH_2{-}C(=O){-}OH \text{ (Hippuric acid (N-benzoylglycine))}$$

Figure 8.10. Metabolic oxidation of toluene with conjugation to hippuric acid, which is excreted with urine.

8.6. NAPHTHALENE

Naphthalene, also known as tar camphor, and its alkyl derivatives, such as 1-(2-propyl)naphthalene (Figure 8.11), are important industrial chemicals. Used to make mothballs, naphthalene is a volatile white crystalline solid with a characteristic odor. Coal tar and petroleum are the major sources of naphthalene. Numerous industrial chemical derivatives are manufactured from it. The most important of these is phthalic anhydride (Figure 8.11), used to make phthalic acid plasticizers, which are discussed in Chapter 9.

Metabolism of Naphthalene

The metabolism of naphthalene is similar to that of benzene, starting with an enzymatic epoxidation of the aromatic ring:

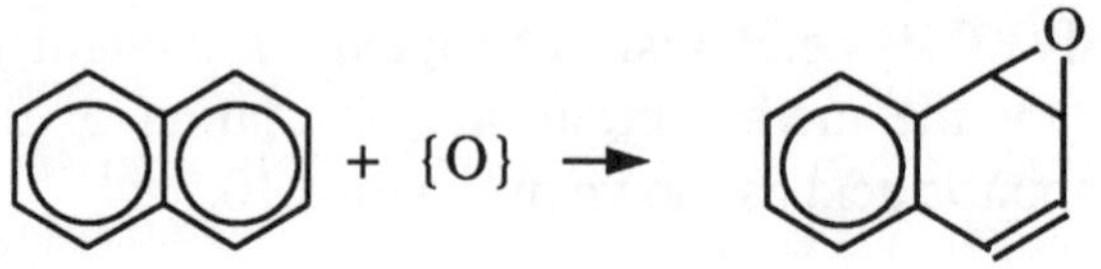

+ {O} → (8.11)

followed by a non-enzymatic rearrangement to 1-naphthol:

→ (8.12)

or addition of water to produce naphthalene-l,2-dihydrodiol through the action of epoxide hydrase enzyme:

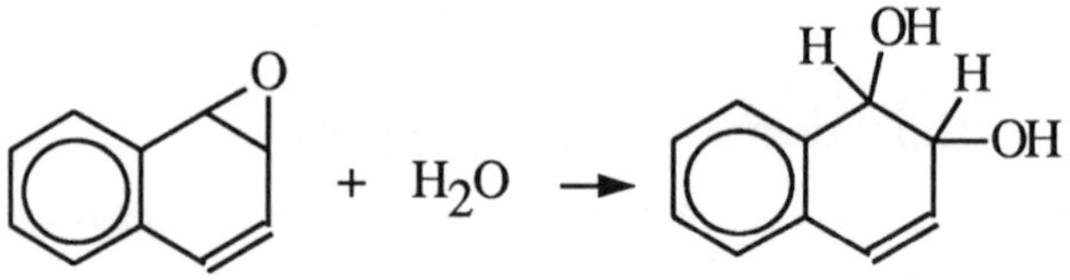

+ H_2O → (8.13)

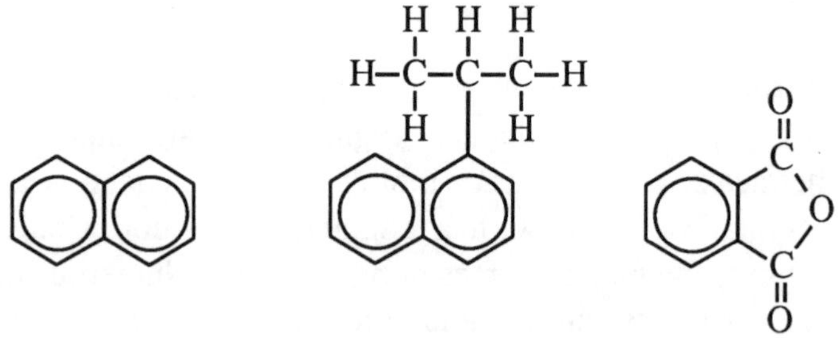

Naphthalene 1–(2–Propyl)naphthalene Phthalic anhydride

Figure 8.11. Naphthalene and two of its derivatives.

Elimination of the metabolized naphthalene from the body may occur as a mercapturic acid, preceded by the glutathione S-transferase-catalyzed formation of a glutathione conjugate.

Toxic Effects of Naphthalene

Exposure to naphthalene can cause a severe hemolytic crisis in some individuals with a genetically-linked metabolic defect associated with insufficient activity of the glucose-6-phosphate dehydrogenase enzyme in red blood cells.[13] Effects include anemia and marked reductions in red cell count, hemoglobin, and hematocrit. Contact of naphthalene with skin can result in skin irritation or severe dermatitis in sensitized individuals. In addition to the hemolytic effects just noted, both inhalation and ingestion of naphthalene can cause headaches, confusion, and vomiting. Kidney failure is usually the ultimate cause of death in cases of fatal poisonings.

8.7. POLYCYCLIC AROMATIC HYDROCARBONS

Benzo(a)pyrene (Figure 8.1) is the most studied of the polycyclic aromatic hydrocarbons (PAHs). These compounds are formed by the incomplete combustion of other hydrocarbons so that hydrogen is consumed in the preferential formation of H_2O. The condensed aromatic ring system of the PAH compounds produced is the thermodynamically favored form of the hydrogen-deficient, carbon-rich residue. To cite an extreme example, the H:C ratio in methane (CH_4) is 4:1, whereas in benzo(a)pyrene ($C_{20}H_{12}$) it is only 3:5.

There are many conditions of partial combustion and pyrolysis that favor production of PAH compounds, and they are encountered abundantly in the atmosphere, soil, and elsewhere in the environment. Sources of PAH compounds include engine exhausts, wood stove smoke, cigarette smoke and char-broiled food. Coal tars and petroleum residues have high levels of PAHs.

PAH Metabolism

The metabolism of PAH compounds is mentioned here with benzo(a)pyrene as an example. Several steps lead to the formation of the carcinogenic metabolite product of benzo(a)pyrene.[11] After an initial oxidation to form the 7,8-epoxide, the 7,8-diol is produced through the action of epoxide hydrase enzyme as shown by the following reaction:

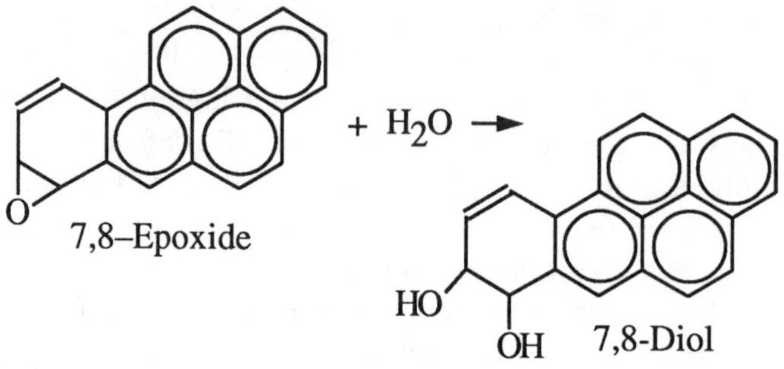

7,8–Epoxide + H_2O → 7,8-Diol (8.14)

The microsomal mixed-function oxidase enzyme system further oxidizes the diol to the carcinogenic 7,8-diol-9,10-epoxide:

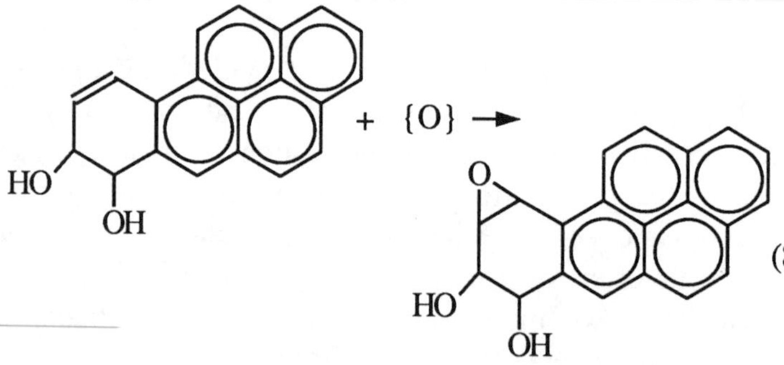

+ {O} → (8.15)

7,8–Diol–9,10–epoxide of benzo(a)pyrene

Toxicology of PAH Compounds

The most notable toxicologic characteristic of PAH compounds is that some of their metabolites are known to cause cancer. The most studied of these is the 7,8-diol-9,10 epoxide of benzo(a)pyrene shown above. There are two stereoisomers of this metabolite, both of which are known to be potent mutagens.

LITERATURE CITED

1. Sax, N. Irving, *Dangerous Properties of Industrial Materials*, 6th ed., Van Nostrand Reinhold, New York, 1984.

2. *Threshold Limit Values for Chemical Substances and Physical Agents in the Work Environment*, American Conference of Governmental Industrial Hygienists (ACGIH), Cincinnati, Ohio, 1983.

3. Garrett, Jack L., Lewis J. Cralley, and Lester V. Cralley, Eds., *Industrial Hygiene Management*, John Wiley and Sons, Inc., Somerset, NJ (1988).

4. Campbell, M. L., "Hydrocarbons — Cyclohexane," in *Kirk-Othmer Concise Encyclopedia of Chemical Technology*, Wiley-Interscience, New York, 1985, p. 621.

5. Cornish, Herbert H., "Solvents and Vapors," Chapter 18 in *Casarett and Doull's Toxicology*, 2nd ed., John Doull, Curtis D. Klaassen and Mary O. Amdur, Eds., Macmillan Publishing Co., New York, 1980.

6. Hueper, W. C., "Medicolegal Considerations of Occupational and Nonoccupational Cancers," *Lawyer's Medical Cyclopedia*, Vol. 5B, C. J. Frankel and R. M. Paterson, Eds., The Allen Smith Co., Indianapolis, Indiana, 1972.

7. Kniel, Ludwig, Olaf Winter, and Chung-Hu Tsai, "Ethylene," in *Kirk-Othmer Concise Encyclopedia of Chemical Technology*, Wiley-Interscience, New York, 1985, pp. 437-439.

8. Layman, Patricia, "Demand for Alpha-Olefins Forecast to Double by Year 2000," *Chemical and Engineering News*, May 30, 1988, American Chemical Society, Washington, DC, pp. 9-10.

9. Kamrin, Michael A. "The Case of Benzene," Chapter 11 in *Toxicology—A Primer on Toxicology Principles and Applications*, Lewis Publishers, Inc, Chelsea, Michigan, 1988, pp. 95-100.

10. Dreisbach, Robert H., and William O. Robertson, *Handbook of Poisoning*, 12th ed., Appleton and Lange, Norwalk, Conn., 1987.

11. Hodgson, Ernest, and Walter C. Dauterman, "Metabolism of Toxicants, Phase I Reactions," Chapter 4 in *Introduction to Biochemical Toxicology*, Ernest Hodgson and Frank E., Guthrie, Eds., Elsevier, New York, 1980, pp. 68–91.

12. Dauterman, Walter C. "Metabolism of Toxicants, Phase II Reactions," Chapter 5 in *Introduction to Biochemical Toxicology*, Ernest Hodgson and Frank E.,Guthrie, Eds., Elsevier, New York, 1980, pp. 92–105.

13. Gosselin, Robert E., Roger P. Smith, and Harold C. Hodge, "Naphthalene," in *Clinical Toxicology of Commercial Products*, 5th ed., Williams and Wilkins, Baltimore/London, 1984, pp. III-307–III-311.

9

Oxygen-Containing Organic Compounds

9.1. INTRODUCTION

A very large number of organic compounds and natural products, many of which are toxic, contain oxygen in their structures. This chapter concentrates on organic compounds that have oxygen covalently bonded to carbon. Organic compounds in which oxygen is bonded to nitrogen, sulfur, phosphorus, and the halogens are discussed in Chapters 10–13.

Oxygen-Containing Functional Groups

As shown in Table 2.1 and Figure 9.1, there are several kinds of oxygen-containing functional groups in organic compounds. In general, the organo-oxygen compounds can be classified according to the degree of oxygenation, location of oxygen on the hydrocarbon moiety, presence of unsaturated bonds in the hydrocarbon structure, and presence or absence of aromatic rings.

Some of the features of organo-oxygen compounds listed above can be seen from an examination of some of the oxidation products of propane in Figure 9.1. The degree of oxygenation increases in the order alcohols–ethers–epoxides<aldehydes–ketones<carboxylic acids.

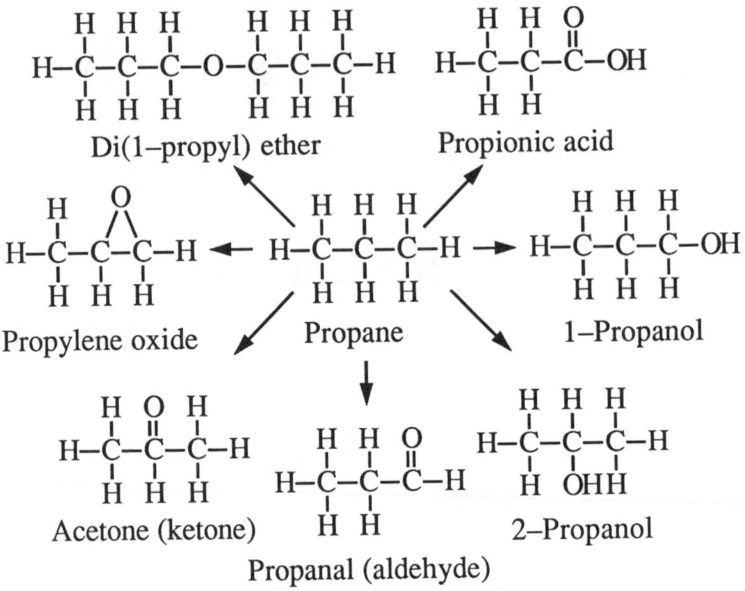

Di(1–propyl) ether
Propionic acid
Propylene oxide
Propane
1–Propanol
Acetone (ketone)
Propanal (aldehyde)
2–Propanol

Figure 9.1. Oxygenated derivatives of propane.

9.2 ALCOHOLS

This section discusses the toxicological chemistry of the **alcohols**, oxygenated compounds in which the hydroxyl functional group is attached to an aliphatic or olefinic hydrocarbon skeleton. The phenols, which have –OH bonded to an aromatic ring, are covered in Section 9.3. The three lightest alcohols — methanol, ethanol, and ethylene glycol (shown in Figure 9.2) — are discussed individually in some detail because of their widespread use and human exposure to them. The higher alcohols, defined broadly as those containing 3 or more carbon atoms per molecule, are discussed as a group.

Methanol
Ethanol
Ethylene glycol

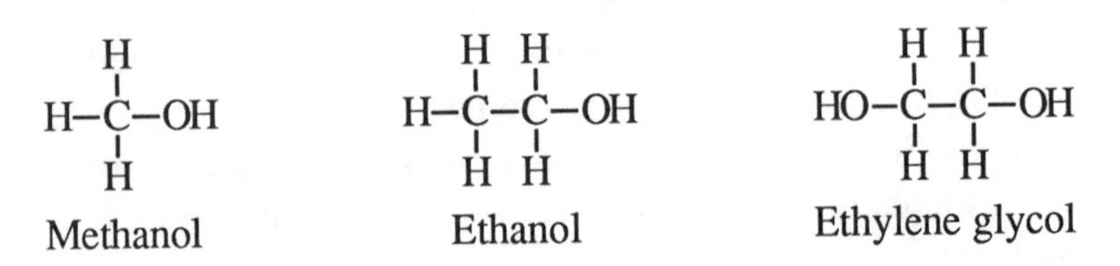

Figure 9.2. Three lighter alcohols with particular toxicological significance.

Methanol

Methanol, also called methyl alcohol or wood alcohol, is a clear liquid (mp -98°C, bp 65°C). Until the early 1900s, the major commercial source of methanol was the destructive distillation of wood, a process that yields a product contaminated with allyl alcohol, acetone, and acetic acid. Now methanol is synthesized by the following reaction of hydrogen gas and carbon monoxide, both readily obtained from natural gas or from coal gasification:

$$CO + 2H_2 \xrightarrow[\text{catalyst}]{\text{Metal}} CH_3OH \qquad (9.1)$$

The greatest use for methanol is in the manufacture of formaldehyde (see Section 9.5). Additional uses include the synthesis of other chemicals, including acetic acid, applications as an organic solvent, and addition to unleaded gasoline for fuel, antifreeze, and antiknock properties. (Methanol can be used as an "oxygenated additive" to gasoline to raise octane rating and reduce carbon monoxide emissions. For the first time in a major area-wide effort to curtail CO emissions by altering motor fuel composition, oxygenated additives for gasoline were required in Denver, Colorado, during the winter of 1987–1988.)

Methanol has been responsible for the deaths of many humans who ingested it accidentally or as a substitute for beverage ethanol. The fatal human dose is believed to lie between 50 and 250 g.[1] In the body, methanol undergoes metabolic oxidation to formaldehyde and formic acid:

$$H_3C\text{–}OH \xrightarrow{\{O\}} H\text{–}C(=O)\text{–}H + H_2O \quad \text{(Formaldehyde)}$$

$$H_3C\text{–}OH \xrightarrow{2\{O\}} H\text{–}C(=O)\text{–}OH + H_2O \quad \text{(Formic acid)} \qquad (9.2)$$

The products of this reaction cause acidosis, but the main effects are on the central nervous system and the optic nerve. In cases of acute exposure, an initially mild inebriation is followed in about 10–20 hours by unconsciousness and cardiac depression, and death may occur. For sublethal doses, initial symptoms of visual dysfunction may clear temporarily followed by blindness from deterioration of the optic nerve and retinal ganglion cells. Chronic exposures to lower levels of methanol may result from fume inhalation.

Ethanol

Ethanol (ethyl alcohol, mp –114°C, bp 78°C) is a clear, colorless liquid widely used as a beverage ingredient, synthetic chemical, solvent, germicide, antifreeze, and gasoline additive. It is produced by the fermentation of carbohydrates or by the hydration of ethylene as shown by the two following reactions:

$$C_6H_{12}O_6 \xrightarrow{\text{Yeasts}} 2C_2H_5OH + 2CO_2 \qquad (9.3)$$

$$H_2C{=}CH_2 + H_2O \xrightarrow{\text{Mixed–bed catalyst}} H{-}\overset{H}{\underset{H}{C}}{-}\overset{H}{\underset{H}{C}}{-}OH \qquad (9.4)$$

Ethanol misused in beverages is responsible for more deaths than any other chemical when account is taken of chronic alcoholism, vehicle fatalities caused by intoxicated drivers, and alcohol-related homicides. Chronic alcoholism is a distinct disease arising from generally long-term systemic effects of the ingestion of alcohol and will not be discussed further here.

Ethanol has a range of acute effects, normally expressed as a function of percent blood ethanol.[2] In general these effects are related to central nervous system depression. Mild effects such as decreased inhibitions and slowed reaction times begin to appear at about 0.05% blood ethanol. Most individuals are clinically intoxicated at a level of 0.15–0.3% blood ethanol, in the 0.3–0.5% range stupor may be produced, and at 0.5% and above coma and often death occur.

Metabolically, ethanol is oxidized first to acetaldehyde (Section 9.6), then to CO_2. The overall oxidation rate is faster than for methanol.

In addition to absorption through the gastrointestinal tract, ethanol can be absorbed by the alveoli of the lungs. Symptoms of intoxication can be observed from inhalation of air containing more than 1000 ppm ethanol.

Ethylene Glycol

Although used in cosmetics, chemical synthesis, and other applications, most ethylene glycol is consumed as the major ingredient of antifreeze/antiboil formulations for automobile radiators. Ethylene glycol (mp -13°C, bp 198°C) is synthesized by the oxidation of ethylene to ethylene oxide, followed by hydrolysis of the latter compound:

$$H_2C{=}CH_2 + \{O\} \rightarrow H{-}\underset{H}{C}{-}\overset{O}{\triangle}{-}\underset{H}{C}{-}H \qquad (9.5)$$

$$H{-}\underset{H}{C}{-}\overset{O}{\triangle}{-}\underset{H}{C}{-}H + H_2O \rightarrow HO{-}\overset{H}{\underset{H}{C}}{-}\overset{H}{\underset{H}{C}}{-}OH \qquad (9.6)$$

Toxic exposures to ethylene glycol are rare because of its low vapor pressure, but inhalation of droplets can be very dangerous. About 50 human fatalities attributable to ethylene glycol poisoning have been documented.[3] From the limited amount of information available, the LD_{50} for humans has been estimated at about 110 g. Ingested ethylene glycol initially stimulates the central nervous system, then depresses it. Victims may suffer acedemia from the presence of the intermediate metabolite, glycolic acid. Kidney damage occurs later, and it can be fatal. The kidneys are harmed because of the deposition of solid calcium oxalate resulting from the following overall process:

$$\underset{\text{Ethylene glycol}}{H{-}\overset{H}{\underset{OH}{C}}{-}\overset{H}{\underset{OH}{C}}{-}H} \xrightarrow[\text{Metabolic processes}]{\{O\}} \text{Oxalate} \xrightarrow{Ca^{2+}} \underset{\text{Calcium oxalate (solid)}}{CaC_2O_4} \qquad (9.7)$$

Important intermediates in this process are glycoaldehyde, glycolate, and glyoxalate.[4] Kidney failure from the metabolic formation of calcium oxalate has been especially common in cat species, which have voracious appetites for ethylene glycol in antifreeze. Deposits of solid calcium oxalate have been observed as well in the liver and brain tissue of victims of ethylene glycol poisoning.

The Higher Alcohols

Numerous alcohols containing three or more carbon atoms are used as solvents and chemical intermediates and for other purposes. Exposure to these compounds can occur and their toxicities are important. Some of the more significant of these alcohols are listed in Table 9.1.

Table 9.1. Some Alcohols with Three or More Carbons

Alcohol name and formula	Properties
2-Propanol, $CH_3CHOHCH_3$	Isopropyl alcohol, used as "rubbing alcohol" and food additive, irritant, narcotic, relatively low toxicity
Allyl alcohol, $CH_2{=}CHCH_2OH$	Olefinic alcohol, pungent odor, strongly irritating to eyes, mouth, lungs
1-Butanol, $CH_3(CH_2)_2CH_2OH$	Butyl alcohol or *n*–butanol, irritant, limited toxicity because of low vapor pressure
1–Pentanol, $CH_3(CH_2)_3CH_2OH$	Amyl alcohol, liquid, bp 138°C, irritant, causes headache and nausea, low vapor pressure and low water solubility reduce toxicity hazard
1–Decanol, $CH_3(CH_2)_8CH_2OH$	Viscous liquid, bp 233°C, low acute toxicity
2-Ethylhexanol, $CH_3(CH_2)_3CH$-$(C_2H_5)CH_2OH$	2–Ethylhexyl alcohol, important industrial solvent, toxicity similar to those of butyl alcohols

An important alcohol in toxicology studies is ***n*-octanol**, $CH_3(CH_2)_6CH_2OH$. This compound is applied to the measurement of the **octanol-water partition coefficient**, which is used to estimate how readily organic toxicants are transferred from water to lipids, a tendency usually associated with ability to cross cell membranes and cause toxic effects. As just one example, the octanol-water partition coefficient can be used to estimate the tendency of organic compounds to be taken up from water to the lipid gill tissue of fish.[5]

9.3. PHENOLS

Phenols are aryl alcohols in which the –OH group is bonded to an aromatic hydrocarbon moiety. The simplest of these compounds is phenol, in which the hydrocarbon portion is the phenyl group. Figure 9.3 shows some of the more important phenolic compounds. Phenols have properties that are quite different from those of the aliphatic and olefinic alcohols. Many important phenolic compounds have nitro groups ($-NO_2$) and halogen atoms (particularly Cl) bonded to the aromatic rings. These substituents may strongly affect chemical and toxicological behavior; such compounds are discussed in Chapters 10 and 11.

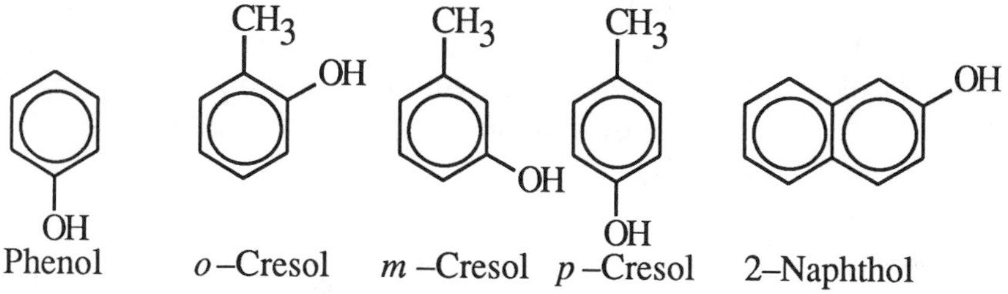

Phenol *o*–Cresol *m*–Cresol *p*–Cresol 2–Naphthol

Figure 9.3. Major phenolic compounds.

Properties and Uses of Phenols

The physical properties of the phenols listed in Figure 9.3 are summarized briefly in Table 9.2. These phenolic compounds are weak acids that ionize to phenolate ions in the presence of base:

$$C_6H_5{-}OH + OH^- \rightarrow C_6H_5{-}O^- + H_2O \quad (9.8)$$

The water solubility of the phenolate salts enables extraction of phenols from coal tar into aqueous base, which is a commercial source of phenolic compounds. The major commercial use of phenol is in the manufacture of phenolic resin polymers, usually with formaldehyde. Phenols and cresols are used as antiseptics and disinfectants in areas such as barns where the phenol odor can be tolerated. Phenol was the original antiseptic used on wounds and in surgery, starting with the work of Lord Lister in 1885.

Toxicology of Phenols

In a general sense, the phenols have similar toxicological effects. Phenol is a protoplasmic poison, so it damages all kinds of cells. Early medical studies that demonstrated the concept of aesepsis with phenol revealed its toxicity as well. Phenol is alleged to have caused "an astonishing number of poisonings"[6] since it came into general use.

Table 9.2. Properties of Major Phenolic Compounds

Compound	Properties
Phenol	Carbolic acid, white solid, characteristic odor, mp 41°C, bp 182°C
m–Cresol	Often occurs mixed with *ortho*- and *para*- analogs as cresol or cresylic acid, light yellow liquid, mp 11°C, bp 203°C
o–Cresol	Solid, mp 31°C, bp 191°C
p–Cresol	Crystalline solid with phenolic odor, mp 36°C, bp 202°C
1–Naphthol	Alpha-naphthol, colorless solid, mp 96°C, bp 282°C
2–Naphthol	Beta-naphthol, mp 122°C, bp 288°C

Fatal doses of phenol may be absorbed through the skin. Its acute toxicological effects are predominantly upon the central nervous system. Death can occur as early as one-half hour after exposure. Key organs damaged by chronic exposure to phenol include the spleen, pancreas, and kidneys. Lung edema can also occur.

Some phenol is eliminated from the body as the unchanged molecular compound, although most is metabolized prior to excretion. As noted in Chapter 4, Phase II reactions in the body result in the conjugation of phenol to produce sulfates and glucuronides. These water-soluble metabolic products are eliminated via the kidneys.

Oral doses of naphthols can be fatal. Acute poisoning by these compounds can cause severe gastrointestinal disturbances, kidney malfunction, circulatory system failure, and convulsions. Naphthols can be absorbed through the skin, one effect of which can be eye damage involving the cornea and lens.

9.4. OXIDES

Hydrocarbon **oxides** are significant for both their uses and their toxic effects.[7] As shown for ethylene oxide (1,2-epoxyethane) in Reactions 9.5 and 9.6 and propylene oxide (1,2-epoxypropane) in Figure 9.1, these compounds are characterized by an **epoxide** functional group consisting of an oxygen atom bridging between two adjacent C atoms. As discussed in Section 4.2, the metabolic formation of such a group is called epoxidation and is a major type of the Phase I reactions of xenobiotic compounds. In addition to ethylene and propylene oxides, four other common hydrocarbon oxides are shown in Figure 9.4.

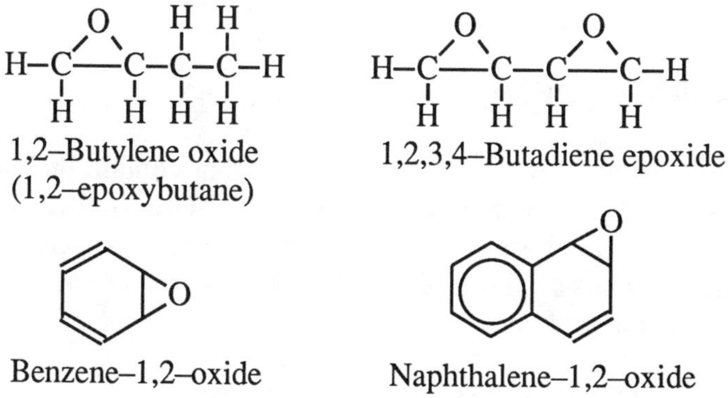

1,2–Butylene oxide (1,2–epoxybutane)

1,2,3,4–Butadiene epoxide

Benzene–1,2–oxide

Naphthalene–1,2–oxide

Figure 9.4. Some common epoxide compounds.

Ethylene oxide (mp -111°C, bp 11°C) is a colorless, sweet-smelling, flammable, explosive gas. It is used as a chemical intermediate, sterilant, and fumigant. It has a moderate to high toxicity, is a mutagen, and is carcinogenic to experimental animals. When inhaled, ethylene oxide causes respiratory tract irritation, headache, drowsiness, and dyspnea. At higher levels, cyanosis, pulmonary edema, kidney damage, peripheral nerve damage, and death can result from inhalation of this compound.

Propylene oxide (mp -104°C, bp 34°C) is a colorless, reactive, volatile liquid with uses similar to those of ethylene oxide. Its toxic effects are like those of ethylene oxide, though less severe. The properties of butylene oxide (liquid, bp 63°C) are also similar to those of ethylene oxide. The oxidation product of 1,3-butadiene, 1,2,3,4-butadiene, is a direct–acting (primary) carcinogen.

As discussed in Section 4.2, benzene-1,2-oxide is an intermediate in the biochemical oxidation of benzene. It is probably responsible for the toxicity of benzene. It is hydrolyzed by the action of epoxide hydratase to the dihydrodiol shown below:

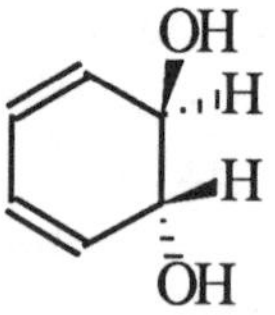

OH
H
H
OH
Benzene *Trans* –1,2–dihydrodiol

Naphthalene-1,2-oxide is a metabolic intermediate in the oxidation of naphthalene mediated by cytochrome P-450.

9.5. FORMALDEHYDE

Aldehydes and ketones are compounds that contain the carbonyl (C=O) group. Of these compounds, **formaldehyde** (shown below) is uniquely important because of its widespread use and toxicity, and it is discussed in this section. Other aldehydes and ketones are discussed in the following section.

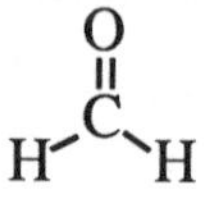

O
C
H H
Formaldehyde

Properties and Uses of Formaldehyde

Formaldehyde (mp -118°C, bp -19°C) is a colorless gas with a pungent, suffocating odor. It is manufactured by the oxidation of methanol over a silver catalyst. Because it undergoes a number of important reactions in chemical synthesis and can be made at relatively low cost, formaldehyde is one of the most widely used industrial chemicals. In the pure form it polymerizes with itself to give hydroxyl compounds, ketones, and other aldehydes. Because of this tendency, commercial formaldehyde is marketed as a 37–50% aqueous solution containing some methanol called **formalin**. Formaldehyde is a synthesis intermediate in the production of resins (particularly phenolic resins), as well as a large number of synthetic organic compounds, such as chelating agents. Formalin is employed in antiseptics, fumigants, tissue and biological specimen preservatives, and embalming fluid.

Toxicity of Formaldehyde and Formalin

It has been noted[8] that formaldehyde is produced by natural processes in the environment and in the body, which would suggest that it might not be very toxic. However, such is not the case in that formaldehyde exhibits a number of toxic effects.

Exposure to inhaled formaldehyde via the respiratory tract is usually to molecular formaldehyde vapor, whereas exposure by other routes is usually to formalin. Exposure to formaldehyde vapor can occur in industrial settings. In recent years a great deal of concern has arisen over the potential for exposure in buildings from formaldehyde vapor evolved from insulating foams that were not properly formulated and cured or when these foams burn. Hypersensitivity can result from prolonged, continuous exposure to formaldehyde. Furthermore, animal experiments have shown formaldehyde to be a lung carcinogen.[9]

The human LD_{50} for the ingestion of formalin has been estimated at around 45 g. Deaths have been caused by as little as about 30 g and individuals have survived ingestion of about 120 g, although in at least one such case removal of the stomach was required. Ingestion results in violent gastrointestinal disturbances, including vomiting and diarrhea. Formaldehyde attacks the mucous membrane linings of both the respiratory and alimentary tracts and reacts strongly with functional groups in molecules. Metabolically, formaldehyde is rapidly oxidized to formic acid (see Section 9.7), which is responsible in large part for its toxicity.[2]

9.6. ALDEHYDES AND KETONES

In **aldehydes** the carbonyl group is attached to a C and H atom at the end of a hydrocarbon chain, and in a **ketone** it is bonded to two C atoms in the middle of a hydrocarbon chain or ring. The hydrocarbon portion of aldehydes and ketones may consist of saturated or unsaturated straight chains, branched chains, or rings. The structures of some important aldehydes and ketones are shown in Figure 9.5.

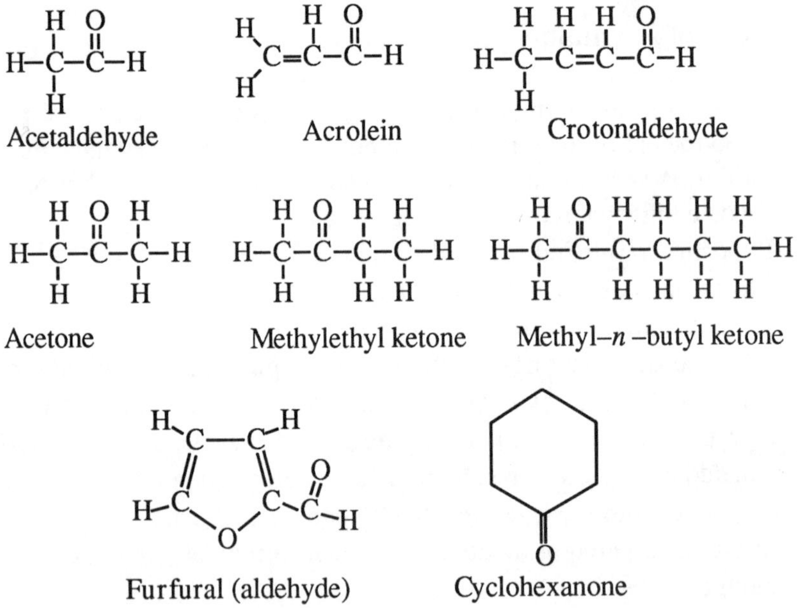

Acetaldehyde Acrolein Crotonaldehyde

Acetone Methylethyl ketone Methyl–*n*–butyl ketone

Furfural (aldehyde) Cyclohexanone

Figure 9.5. Aldehydes and ketones that are significant for their commercial uses and toxicological importance.

Both aldehydes and ketones are industrially important classes of chemicals. Aldehydes are reduced to make the corresponding alcohols and are used in the manufacture of resins, dyes, plasticizers, and alcohols. Some aldehydes are ingredients in perfumes and flavors. Several ketones are excellent solvents and are widely used for that purpose to dissolve gums, resins, laquers, nitrocellulose, and other substances.

Toxicities of Aldehydes and Ketones

In general, because of their water solubility and intensely irritating qualities, the lower aldehydes attack exposed moist tissue, particularly the eyes and mucous membranes of the upper respiratory tract. Because of their lower water solubility, the lower aldehydes can penetrate further into the respiratory tract and affect the lungs.[10]

The toxicity of formaldehyde was discussed in the preceding section. Acetaldehyde is a colorless liquid (bp 21°C) and acts as an irritant and systemically as a narcotic to the central nervous system. Acrolein, an olefinic aldehyde, is a colorless to light yellow liquid (bp 52°C). It is a very reactive chemical that polymerizes readily. It is quite toxic by all routes of contact and ingestion. It has a choking odor and is extremely irritating to respiratory tract membranes. It is classified as an extreme lachrymator (substance that causes eyes to water). Because of this property, acrolein serves to warn of its own exposure. It can produce tissue necrosis, and direct contact with the eye can be especially hazardous.[11] Crotonaldehyde is similarly dangerous and can cause burns to the eye cornea.

Metabolically, aldehydes are converted to the corresponding organic acids as shown by the following general reaction:

$$R-\overset{\overset{\displaystyle O}{\|}}{C}-H + \{O\} \longrightarrow R-\overset{\overset{\displaystyle O}{\|}}{C}-OH \qquad (9.9)$$

In mammals, the liver enzymes aldehyde dehydrogenase and aldehyde oxidase appear to be mainly responsible for this reaction.[12]

Acetone is a liquid with a pleasant odor. It can act as a narcotic and dissolves fats from skin, causing dermatitis. Methyl-*n*-butyl ketone, a widely used solvent, is a mild neurotoxin. Methylethyl ketone is suspected of having caused neuropathic disorders in shoe factory workers.

9.7. CARBOXYLIC ACIDS

Carboxylic acids contain the –C(O)OH functional group bound to an aliphatic, olefinic, or aromatic hydrocarbon moiety. This section deals with those carboxylic acids that contain only C, H, and O. Carboxylic acids that contain other elements, such as trichloroacetic acid (a strong acid) or deadly poisonous monofluoroacetic acid, are discussed in later chapters. Some of the more significant carboxylic acids are shown in Figure 9.6.

Carboxylic acids are the oxidation products of aldehydes and are often synthesized by that route. Some of the higher carboxylic acids are constituents of oil, fat, and wax esters[13] from which they are prepared by hydrolysis. Carboxylic acids have many applications. Formic acid is used as a relatively inexpensive acid to neutralize base, in the treatment of textiles, and as a reducing agent. Acetic and propionic acids are added to foods for flavor and as preservatives. Among numerous other applications, these acids are also used to make cellulose plastics. Stearic acid acts as a dispersive agent and accelerator activator in rubber manufacture. Sodium stearate is a major ingredient of most soaps. Many preservative and antiseptic formulations contain benzoic acid. Large quantities of phthalic acid are used to make phthalate ester plasticizers (see Section 9.13). Acrylic acid and methacrylic acid (acrylic acid in which the alpha-hydrogen has been replaced with a $-CH_3$ group; see Figure 9.6) are used in large quantities to make acrylic polymers.

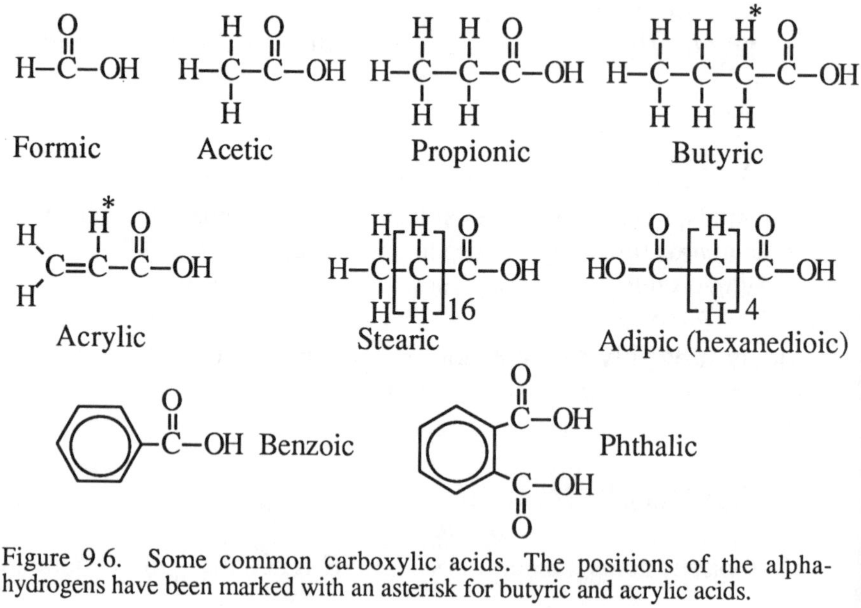

Formic Acetic Propionic Butyric

Acrylic Stearic Adipic (hexanedioic)

Benzoic Phthalic

Figure 9.6. Some common carboxylic acids. The positions of the alpha-hydrogens have been marked with an asterisk for butyric and acrylic acids.

Toxicology of Carboxylic Acids

Concentrated solutions of formic acid are corrosive to tissue, much like strong mineral acids. In Europe decalcifier formulations containing about 75% formic acid have been marketed for removing mineral scale. Children ingesting this material have suffered corrosive lesions to mouth and esophogeal tissue. Although acetic acid is widely used in food preparation as a 4–6% solution in vinegar, pure acetic acid (glacial acetic acid) is extremely corrosive to tissue that it contacts. Acrylic and methacrylic acids are considered to be relatively toxic, both orally and by skin contact. In general, the presence of more than one carboxylic acid group per molecule, unsaturated bonds in the carbon skeleton, or the presence of a hydroxide group on the alpha-carbon position (see Figure 9.6) increases corrosivity and toxicity of carboxylic acids.

9.8. ETHERS

Three important classes of oxygenated organic compounds can be regarded as products of condensation of compounds containing the –OH group accompanied by a loss of H_2O, as shown by the following reaction:

$$R\text{–}OH + HO\text{–}R' \rightarrow R\text{–}O\text{–}R' + H_2O \qquad (9.10)$$

In this reaction, R–OH and HO–R' are either alcohols or carboxylic acids. When both are alcohols, R–O–R' is an ether; when one is an acid and the other an alcohol, the product is an ester; and when both are acids, an acid anhydride is produced. Ethers are discussed in this section, and the other two classes of products are discussed in the two sections that follow.

Examples and Uses of Ethers

An ether consists of two hydrocarbon moieties linked by an oxygen atom as shown in Figure 9.7. Although diethyl ether is highly flammable, ethers are generally not very reactive. This property enables their uses in applications where an unreactive organic solvent is required. Some ethers form explosive peroxides when exposed to air as shown by the example of diethyl ether peroxide in Figure 9.7.

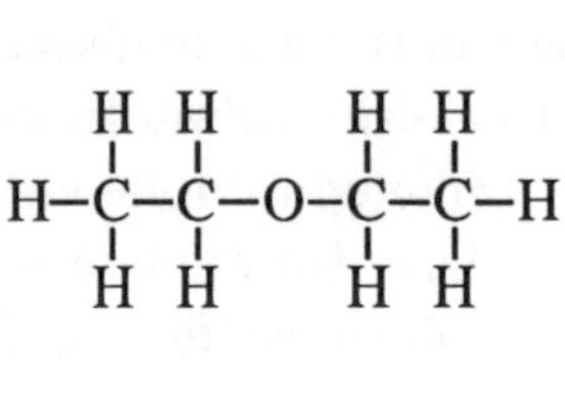

Diethyl ether 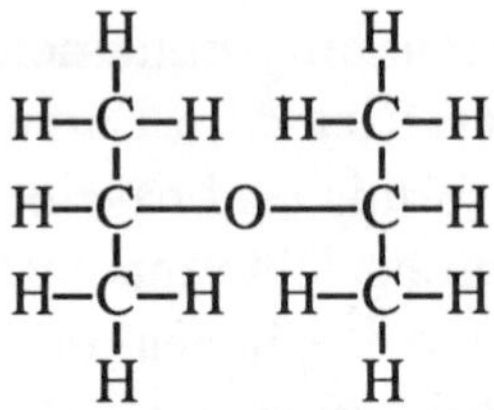 Di-isopropyl ether 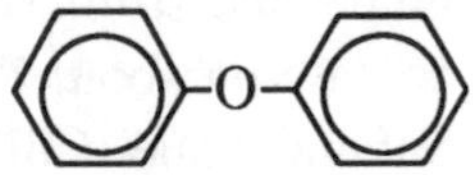 Diphenyl ether

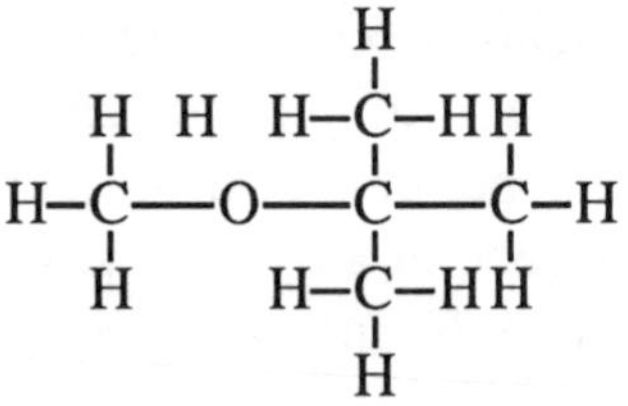

Methyl *tert*- butyl ether 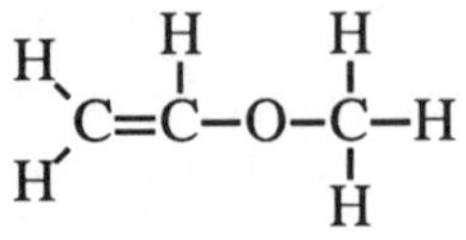 Vinyl methyl ether

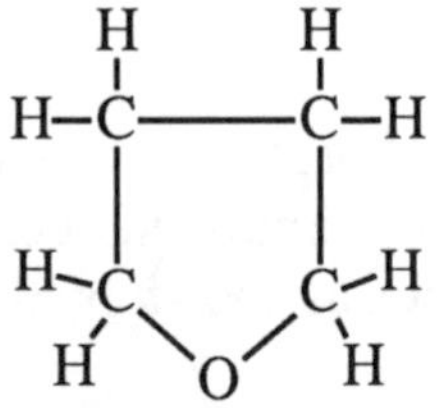

Tetrahydrofuran (cyclic ether) 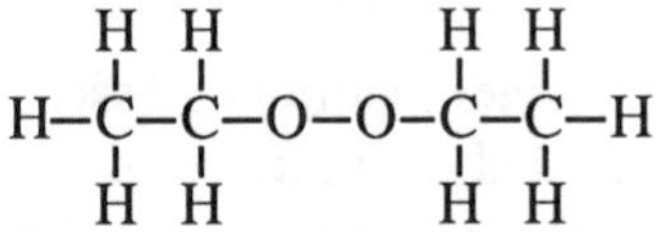 Peroxide from diethyl ether

Figure 9.7. Structures of some common ethers.

Diethyl ether (mp -116°C, bp 34.6°C) is the most commercially important ether. It is used as a reaction medium, solvent, and extractant. The production of methyl *tert*-butyl ether has increased markedly in recent years because of its application as an antiknock ingredient of unleaded gasoline.

Toxicities of Ethers

Because of its volatility, the most likely route of exposure to diethyl ether is by inhalation. About 80% of this compound that gets into the body is eliminated unmetabolized as the vapor through the lungs. Diethyl ether is a central nervous system depressant, and for many years was the anesthetic of choice for surgery. At low doses it causes drowsiness, intoxication, and stupor. Higher exposures result in unconsciousness and even death.

Compared to other classes of organic compounds, the ethers have relatively low toxicities. This characteristic can be attributed to the low reactivity of the C–O–C functional group arising from the high strength of the carbon-oxygen bond. Like diethyl ether, several of the more volatile ethers affect the central nervous system. Hazards other than their toxicities tend to be relatively more important for the ethers. These hazards are flammability and formation of explosive peroxides (especially with di-isopropyl ether).

9.9. ACID ANHYDRIDES

The most important carboxylic **acid anhydride** is acetic anhydride, the structure of which is shown below:

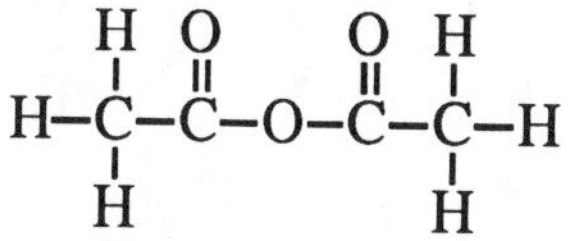

```
  H O   O H
  |  ||  || |
H–C–C–O–C–C–H   Acetic anhydride
  |         |
  H         H
```

Annual world production of this chemical compound is on the order of a million metric tons. In chemical synthesis it functions as an acetylating agent (addition of $CH_3C(O)$ moiety). Its greatest single use is to make cellulose acetate and it has additional applications in manufacturing textile sizing agents, the synthesis of salicylic acid (for aspirin manufacture), electrolytic polishing of aluminum, and the processing of semiconductor components.

Toxicological Considerations

In contrast to the relative safety of many ethers and esters, acetic anhydride is a systemic poison and especially corrosive to the skin, eyes, and upper respiratory tract. Levels in the air of as low as 0.4 mg/m^3 adversely affect eyes, and contamination should be kept to less than 1/10 that level in the workplace atmosphere. Blisters and burns that heal only slowly result from skin exposure. Acetic anhydride has a very strong acetic acid odor that causes an intense burning sensation in the nose and throat that is accompanied by coughing. It is a powerful lachrymator. Fortunately, these unpleasant symptoms elicit a withdrawal response in exposed individuals.

9.10. ESTERS

An ester is formed from an alcohol and acid as shown in Reaction 9.10. Esters exhibit a wide range of biochemical diversity and large numbers of them occur naturally. Fats, oils, and waxes are esters, as are many of the compounds responsible for odors and flavors of fruits, flowers, and other natural products. It follows that many esters are not toxic. Synthetic versions of many of the esters that occur naturally are produced for purposes such as flavoring ingredients. A number of esters that are not natural products have been synthesized for various purposes. Esters are used in industrial applications as solvents, plasticizers, lacquers, soaps, and surfactants. Figure 9.8 shows some representative esters.

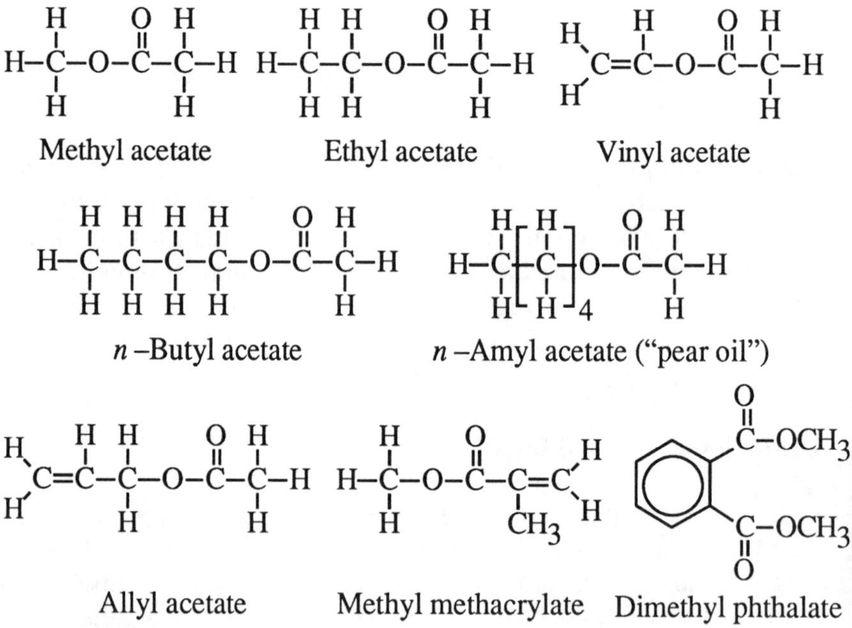

Methyl acetate   Ethyl acetate   Vinyl acetate

n–Butyl acetate   *n*–Amyl acetate ("pear oil")

Allyl acetate   Methyl methacrylate   Dimethyl phthalate

Figure 9.8. Some typical esters.

Methyl acetate is a colorless liquid with a pleasant odor. It is used as a solvent and as an additive to give foods a fruit-like taste. Ethyl acetate is a liquid with a pleasant odor. Liquid vinyl acetate polymerizes when exposed to light to yield a solid polymer. Both *n*-butyl acetate and *n*-amyl acetate are relatively higher-boiling liquids compared to the esters mentioned above. Amyl acetate has a characteristic odor of bananas and pears. Methyl methacrylate is the monomer used to make some kinds of polymers noted for their transparency and resistance to weathering. Among their other applications, these polymers are used as substitutes for glass, particularly in automobile lights. Dimethyl phthalate is the simplest example of the environmentally important phthalate esters. Other significant members of this class of compounds are diethyl, di-*n*-butyl, di-*n*-octyl, bis(2-ethylhexyl), and butyl benzyl phthalates.[14] Used in large quantities as plasticizers to improve the qualities of plastics, these compounds have become widespread environmental pollutants. The higher-molecular-mass phthalate compounds, especially, tend to be environmentally persistent.

Toxicities of Esters

The most common reaction of esters in exposed tissues is hydrolysis:

$$\underset{\text{Ester}}{R{-}O{-}\overset{\overset{\displaystyle O}{\|}}{C}{-}R'} + H_2O \longrightarrow \underset{\text{Alcohol}}{R{-}OH} + \underset{\text{Carboxylic acid}}{H{-}O{-}\overset{\overset{\displaystyle O}{\|}}{C}{-}R'} \qquad (9.13.1)$$

To a large extent, therefore, the toxicities of esters tend to be those of their hydrolysis products. Two physical characteristics of many esters that affect their toxicities are relatively high volatility, which promotes exposure by the pulmonary route, and good solvent action, which affects penetration and tends to dissolve body lipids. Many volatile esters exhibit asphyxiant and narcotic action. As expected for compounds that occur naturally in foods, some esters are non-toxic (in reasonable doses). However, some of the synthetic esters, such as allyl acetate, have relatively high toxicities. As an example of a specific toxic effect, vinyl acetate acts as a skin defatting agent.

Although environmentally persistent, most of the common phthalates have low toxicity ratings of 2 or 3. Phthalate ester toxicities and metabolism have been summarized in a book on the subject.[15]

LITERATURE CITED

1. Gosselin, Robert E., Roger P. Smith, and Harold C. Hodge, "Methanol," in *Clinical Toxicology of Commercial Products*,5th ed., Williams and Wilkins, Baltimore/London, 1984, pp. III-275–III-279.

2. Dreisbach, Robert H., and William O. Robertson, *Handbook of Poisoning*, 12th ed., Appleton and Lange, Norwalk, Conn., 1987.

3. Gosselin, Robert E., Roger P. Smith, and Harold C. Hodge, "Ethylene Glycol," in *Clinical Toxicology of Commercial Products,* 5th ed., Williams and Wilkins, Baltimore/London, 1984, pp. III-172–III-179.

4. Gabow, Patricia A., Keith Clay, John B. Sullivan, and Ronald Lepoff, "Ethylene Glycol Intoxication," *American Journal of Kidney Disease*, **XI**(3), 277-279 (1988).

5. Hodson, Peter V., D. George Dixon, and Klaus L. E. Kaiser, "Estimating The Acute Toxicity of Waterborne Chemicals in Trout from Measurements of Median Lethal Dose and the Octanol-Water Partition Coefficient," *Environmental Toxicology and Chemistry*, **7**, 443–454 (1988).

6. Gosselin, Robert E., Roger P. Smith, and Harold C. Hodge, "Phenol," in *Clinical Toxicology of Commercial Products*, 5th ed., Williams and Wilkins, Baltimore/London, 1984, pp. III-344–III-348.

7. Goldberg, Leon, *Hazard Assessment of Ethylene Oxide*, CRC Press, Boca Raton, Florida, 1986.

8. Kamrin, Michael A., "The Case of Formaldehyde," Chapter 10 in *Toxicology–A Primer on Toxicology Principles and Applications*, Lewis Publishers, Inc, Chelsea, Michigan, 1988, pp. 87–93.

9. *Registry of Toxic Effects of Chemical Substances*, National Institute of Occupational Safety and Health, Washington, D.C., 1976.

10. Sax, N. Irving, *Dangerous Properties of Industrial Materials*, 6th ed., Van Nostrand Reinhold, New York, 1984.

11. Williams, Phillip L., and James L. Burson, *Industrial Toxicology*, Van Nostrand Reinhold Co., New York, 1985.

12. Hodgson, Ernest, and Patricia E. Levi, *Modern Toxicology* Elsevier, New York, 1987.

13. Knuth, C. J.,"Carboxylic Acids," in *The Encyclopedia of Chemistry*, Clifford A. Hampel and Gessner G. Hawley, Eds., Van Nostrand Reinhold Co., New York, 1973, pp. 204–205.

14. "Phthalates," in *Treatability Manual*, EPA 600/8-80/042a, United States Environmental Protection Agency, Washington, D.C., 1980, p. I.6.

15. Woodward, N. Kevin, *Phthalate Esters: Toxicity and Metabolism*, Vol. I and II, CRC Press, Boca Raton, Florida, 1988.

10

Nitrogen-Containing Organic Compounds

10.1. INTRODUCTION

Nitrogen occurs in a wide variety of organic compounds of both synthetic and natural origin. This chapter discusses organic compounds that contain carbon, hydrogen, and nitrogen. Many significant organic nitrogen compounds contain oxygen as well, and these are covered in later parts of the chapter.

10.2. NON-AROMATIC AMINES

Lower Aliphatic Amines

Amines may be regarded as derivatives of ammonia, NH_3, in which 1 to 3 of the H atoms have been replaced by hydrocarbon groups. When these groups are aliphatic groups of which none contains more than 6 C atoms, the compound may be classified as a **lower aliphatic amine**. Among the more commercially important of these amines are mono-, di-, and trimethylamine; mono-, di-, and triethylamine; dipropylamine, isopropylamine, butylamine, dibutylamine, diisobutylamine, cyclohexylamine, and dicyclohexylamine.[1] Example structures are given in Figure 10.1.

The structures in Figure 10.1 indicate some important aspects of amines. Methylamine, methyl-2-propylamine, and triethylamine are primary, secondary, and tertiary amines, respectively. A primary amine has 1 hydrocarbon group substituted for H on NH_3, a secondary amine has 2, and a tertiary amine has 3. Dicyclohexylamine has two cycloalkane substituent groups attached and is a secondary amine.

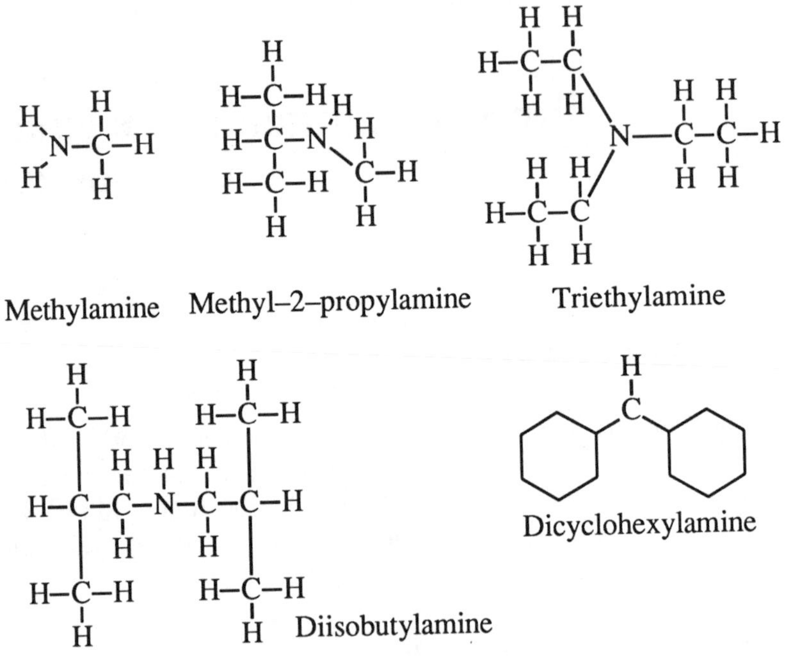

Methylamine Methyl–2–propylamine Triethylamine

Dicyclohexylamine

Diisobutylamine

Figure 10.1. Examples of lower aliphatic amines.

All of the aliphatic amines have strong odors. Of the compounds listed above as commercially important aliphatic amines, the methylamines and monoethylamine are gases under ambient conditions, whereas the others are colorless volatile liquids. The lower aliphatic amines are highly flammable. They are used primarily as intermediates in the manufacture of other chemicals, including polymers (rubber, plastics, textiles), agricultural chemicals, and medicinal chemicals.

The lower aliphatic amines are generally among the more toxic substances in routine, large-scale use. One of the reasons for their toxicity is that they are basic compounds and raise the pH of exposed tissue by hydrolysis with water in tissue as shown by the following reaction:

$$\underset{\text{Amine}}{R_3N} + H_2O \rightarrow R_3NH^+ + OH^- \quad (10.1)$$

Furthermore, these compounds are rapidly and easily taken into the body by all common exposure routes.[2] The lower amines are corrosive to tissue and can cause tissue necrosis at the point of contact. Sensitive eye tissue is vulnerable to amines. These compounds can have systemic effects upon many organs in the body. Necrosis of the liver and kidneys can occur and exposed lungs can exhibit hemorrhage and edema. The immune system may become sensitized to amines.

Of the lower aliphatic amines, cyclohexylamine and dicyclohexylamine appear to have received the most attention for their toxicities. In addition to its caustic effects on eyes, mucous membranes, and skin, cyclohexylamine acts as a systemic poison. In humans the symptoms of systemic poisoning by this compound include nausea to the point of vomiting, anxiety, restlessness, and drowsiness. It adversely affects the female reproductive system. Dicyclohexylamine produces similar symptoms, but is considered to be more toxic. It is appreciably more likely to be absorbed in toxic levels through the skin, probably because of its less polar, more lipid-soluble nature.

Fatty Amines

Fatty amines are those containing alkyl groups having more than 6 carbon atoms. The commercial fatty amines are synthesized from fatty acids that occur in nature and are used as chemical intermediates. Other major uses of fatty amines and their derivatives include textile chemicals (particularly fabric softeners), emulsifiers for petroleum and asphalt, and flotation agents for ores.

Some attention has been given to the toxicity of octadecylamine, which contains a straight-chain, 18-carbon alkane group, because of its use as an anticorrosive agent in steam lines. There is some evidence to suggest that the compound is a primary skin sensitizer.

Alkyl Polyamines

Alkyl polyamines are those in which two or more amino groups are bonded to alkane moieties. The structures of the four most significant of these are shown in Figure 10.2. These compounds have a number of commercial uses, such as for solvents, emulsifiers, epoxy resin hardeners, stabilizers, and starting materials for dye synthesis. They also act as chelating agents; triethylenetetramine is especially effective for that purpose. Largely as a result of their strong alkalinity, the alkyl polyamines tend to be skin, eye, and respiratory tract irritants. The lower homologues are relatively stronger irritants.

Of the common alkyl polyamines, ethylenediamine is the most notable because of its widespread use and toxicity. Although it has a toxicity rating of only 3, it can be very damaging to the eyes and is a strong skin sensitizer. The dihydrochloride and dihydroiodide salts have some uses as human and veterinary pharmaceuticals. The former is administered to acidify urine and the latter as an iodine source. Putrescine is a notoriously odorous naturally-occurring substance produced by bacteria in decaying flesh.

```
H    H H    H
 \   | |   /
  N--C--C--N
 /   | |   \
H    H H    H
Ethylenediamine
```

```
H    H H H H H H H H H H H
 \   | |   | |   | |   | |
  N--C--C--N--C--C--N--C--C--N--C--C--N
 /   | |   | |   | |   | |
H    H H   H H   H H   H H
Tetraethylenepentamine
```

```
H    H H H H H    H
 \   | |   | |   /
  N--C--C--N--C--C--N
 /   | |   | |   \
H    H H   H H    H
Diethylenetriamine
```

```
H    H H H H H H H H    H
 \   | |   | |   | |   /
  N--C--C--N--C--C--N--C--C--N
 /   | |   | |   | |   \
H    H H   H H   H H    H
Triethylenetetramine
```

```
H    H H H H    H
 \   | | | |   /
  N--C--C--C--C--N
 /   | | | |   \
H    H H H H    H
```
Putrescine (odorous product of decayed flesh)

Figure 10.2. Alkyl polyamines in which two or more amino groups are bonded to an alkane group.

Cyclic Amines

Four simple amines in which H atoms are contained in a ring structure are shown in Figure 10.3. The special case of the 6-membered-ring aromatic amine pyridine is discussed in Section 10.4.

Of the compounds shown in Figure 10.3, the first three are liquids under ambient conditions and have the higher toxicity hazards expected of liquid toxicants. All four compounds are colorless in the pure form, but pyrrole darkens upon standing. All are considered to be toxic via the oral, dermal, and inhalation routes. Because of its low volatility, there is little likelihood of inhaling piperazine, except as a dust.

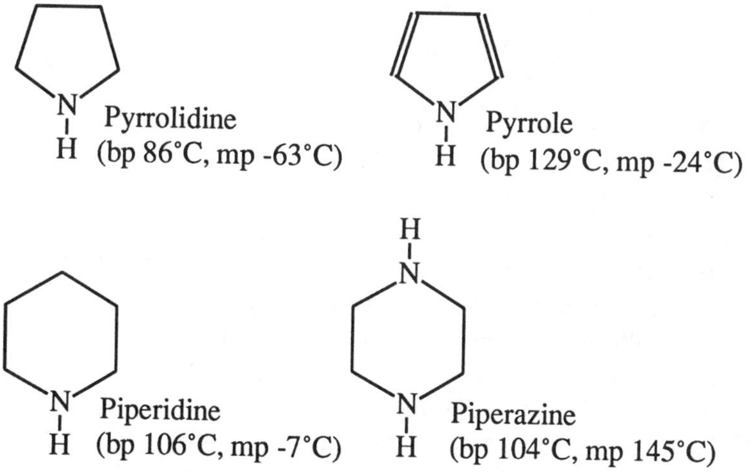

Pyrrolidine (bp 86°C, mp -63°C)

Pyrrole (bp 129°C, mp -24°C)

Piperidine (bp 106°C, mp -7°C)

Piperazine (bp 104°C, mp 145°C)

Figure 10.3. Some common cyclic amines.

10.3. CARBOCYCLIC AROMATIC AMINES

Carbocyclic aromatic amines are those in which at least one substituent group is an aromatic ring containing only C atoms as part of the ring structure, and with one of the C atoms in the ring bonded directly to the amino group. There are numerous compounds with many industrial uses in this class of amines. They are of particular toxicological concern because several have been shown to cause cancer in the human bladder, ureter, and pelvis, and are suspected of being lung, liver, and prostate carcinogens.

Aniline

Aniline (structure below) has been an important industrial chemical for many decades. Currently it is most widely used for the manufacture of polyurethanes and rubber, with lesser amounts consumed in the production of pesticides (herbicides, fungicides, insecticides, animal repellants), defoliants, dyes, antioxidants, antidegradants, and vulcanization accelerators.[3] It is also an ingredient of some household products such as polishes (stove and shoe), paints, varnishes, and marking inks. Aniline is a colorless liquid with an oily consistency and distinct odor, freezing at -6.2°C and boiling at 184.4°C.

$$C_6H_5-NH_2 \quad \text{Aniline}$$

Aniline is considered to be very toxic, with a toxicity rating of 4. It readily enters the body by inhalation, ingestion, and through the skin.[3] In its absorption and toxicological characteristics, aniline resembles nitrobenzene, which is discussed in Section 10.5.

The most common effect of aniline in humans is methemoglobinemia caused by the oxidation of iron(II) in hemoglobin to iron(III) with the result that the hemoglobin can no longer transport oxygen in the body. This condition is characterized by cyanosis and a brown-black color of the blood. Unlike the condition caused by reversible binding of carbon monoxide to hemoglobin, oxygen therapy does not reverse the effects of methemoglobinemia. The effects can be reversed by the action of methemoglobin reductase enzyme as shown by the following reaction:

$$\text{HbFe(III)} \xrightarrow{\text{Methemoglobin reductase}} \text{HbFe(II)} \qquad (10.2)$$

Rodents (mice, rats, rabbits) have a higher activity of this enzyme than do humans, so that extrapolation of rodent experiments with methemoglobinemia to humans is usually inappropriate. Methylene blue can also bring about the reduction of HbFe(III) to HbFe(II) and is used as an antidote for aniline poisoning.

Methemoglobinemia has resulted from exposure to aniline used as a vehicle in indelible laundry-marking inks, particularly those used to mark diapers. This condition was first recognized in 1886, and cases were reported for many decades thereafter. Infants who develop methemoglobinemia from this source suffer a 5–10% mortality rate. The skin of infants is more permeable to aniline than that of adults and infant blood is more susceptible to methemoglobinemia.

Aniline must undergo biotransformation to cause methemoglobinemia because pure aniline does not oxidize iron(II) in hemoglobin to iron(III) in vitro. It is believed that the actual toxic agents formed from aniline are aminophenol and phenyl N-hydroxylamine shown in Figure 10.4. The hepatic detoxification mechanisms for aniline are not very effective. The metabolites of aniline excreted from the body are N-acetyl, N-acetyl-*p*-glucuronide, and N-acetyl-*p*-sulfate products, likewise shown in Figure 10.4.

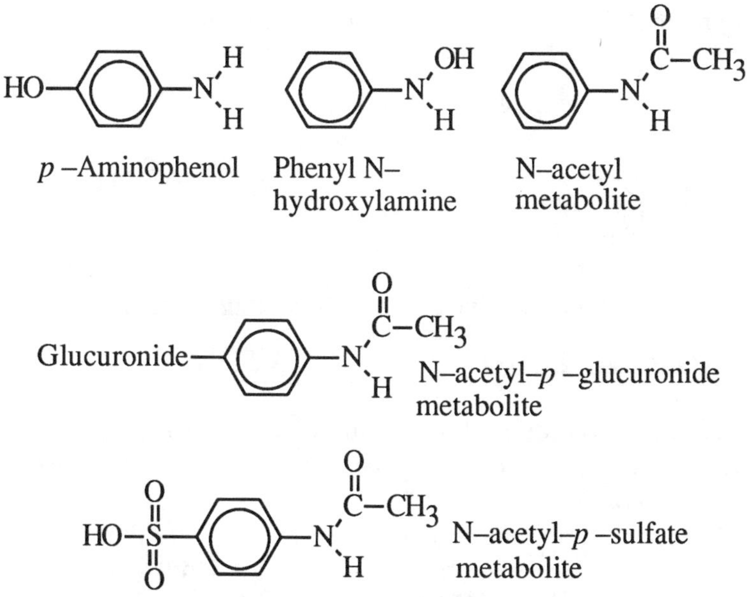

p–Aminophenol

Phenyl N–hydroxylamine

N–acetyl metabolite

N–acetyl–*p*–glucuronide metabolite

N–acetyl–*p*–sulfate metabolite

Figure 10.4. Metabolites of aniline that are toxic or excreted.

Benzidine

Benzidine, *p*-aminodiphenyl, is a solid compound that can be extracted from coal tar. It is highly toxic by oral ingestion, inhalation, and skin sorption and is one of the few proven human carcinogens. Its systemic effects include blood hemolysis, bone marrow depression, and kidney and liver damage.

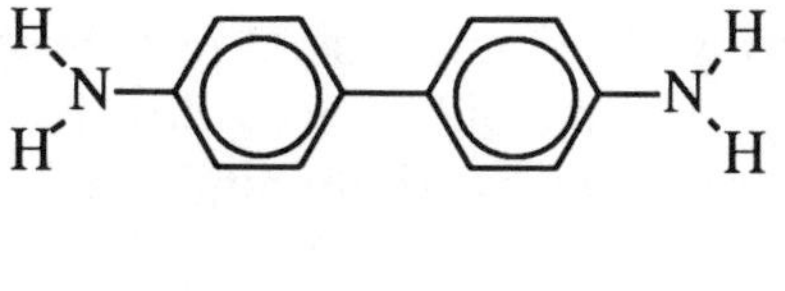

Benzidine

Naphthylamines

The two derivatives of naphthalene having single amino substituent groups are **1-naphthylamine** (alpha-naphthylamine) and **2-naphthylamine** (beta-naphthylamine). Both of these compounds are solids (lump, flake, dust) under normal conditions, although they may be encountered as liquids and vapors. Exposure can occur through inhalation, the gastrointestinal tract, or skin. Both compounds are highly toxic and are proven human bladder carcinogens.

1–Naphthylamine 2–Naphthylamine

10.4. PYRIDINE AND ITS DERIVATIVES

Pyridine is a colorless liquid (mp -42°C, bp 115°C) with a sharp, penetrating odor than can perhaps best be described as "terrible." It is an aromatic compound in which an N atom is part of a 6-membered ring. The most important derivatives of pyridine are the mono-, di-, and trimethyl derivatives; the 2-vinyl and 4-vinyl derivatives; 5-ethyl-2-methylpyridine (MEP); and piperidine (hexahydropyridine, below):

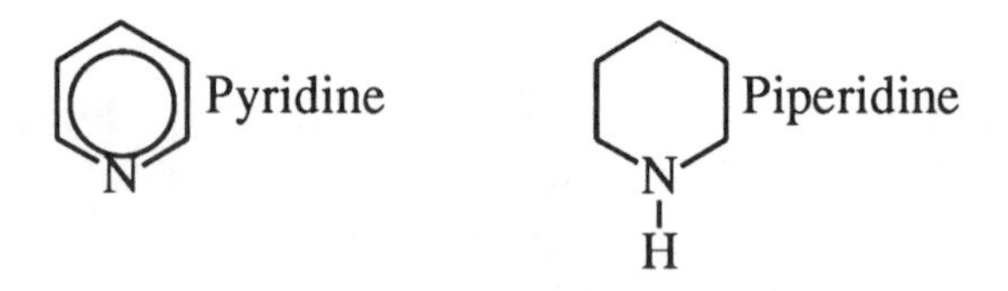

Pyridine and its substituted derivatives are recovered from coal tar. They tend to react like benzene and its analogous derivatives because of the aromatic ring. The major use of pyridine is as an initiator in the process by which rubber is vulcanized. Although considered moderately toxic with a toxicity rating of 3, pyridine has caused fatalities. Symptoms of pyridine poisoning include anorexia, nausea, and fatigue. Its major psychopathological effect is mental depression.

10.5. NITRILES

Nitriles are organic analogs of highly toxic hydrogen cyanide, HCN (see Section 7.2), where the H is replaced by a hydrocarbon moiety. The two most common nitriles are acetonitrile and acrylonitrile, shown below:

$$H-\underset{H}{\overset{H}{C}}-C\equiv N \text{ Acetonitrile} \qquad \begin{matrix} H \\ H \end{matrix}\!\!>C=\overset{H}{C}-C\equiv N \text{ Acrylonitrile}$$

Acetonitrile (mp -45C, bp 81°C) is a colorless liquid with a mild odor. Because of its good solvent properties for many organic and inorganic compounds and its relatively low boiling point, it has numerous industrial uses, particularly as a reaction medium that can be recovered.[4] Acetonitrile has a toxicity rating of 3–4; exposure can occur via the oral, pulmonary, and dermal routes. Although it is considered relatively safe, it is capable of causing human deaths, perhaps by metabolic release of cyanide.

Acrylonitrile is a colorless liquid with a peach-seed odor used in large quantities in the manufacture of acrylic fibers, dyes, and pharmaceutical chemicals. Containing both nitrile and C=C groups, acrylonitrile is a highly reactive compound with a strong tendency to polymerize. It has a toxicity rating of 5 with a mode of toxic action resembling that of HCN. In addition to ingestion, it can be absorbed through the skin or by inhalation of the vapor. It causes blisters and arythema on exposed skin. During metabolic processes acrylonitrile releases cyanide, and its major acute toxic effect is to inhibit enzymes responsible for respiration in tissue, thereby preventing tissue cells from utilizing oxygen. It is a suspect carcinogen.

Acetone cyanohydrin (structure below) is an oxygen-containing nitrile that should be mentioned because of its extreme toxicity and widespread industrial applications. It is used to initiate polymerization reactions and in the synthesis of foaming agents, insecticides, and pharmaceutical compounds. A colorless liquid readily absorbed through the skin, in the body it decomposes to hydrogen cyanide, to which it should be considered toxicologically equivalent (toxicity rating 6) on a molecule-per-molecule basis.

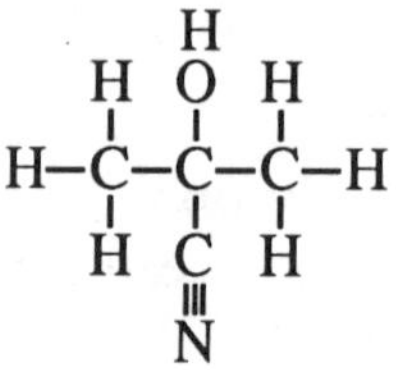

```
        H
    H   O   H
    |   |   |
H—C—C—C—H   Acetone cyanohydrin
    |   |   |
    H   C   H
        |||
        N
```

10.6. NITRO COMPOUNDS

The structures of three significant **nitro compounds**, which contain the $-NO_2$ functional group, are given in Figure 10.5.

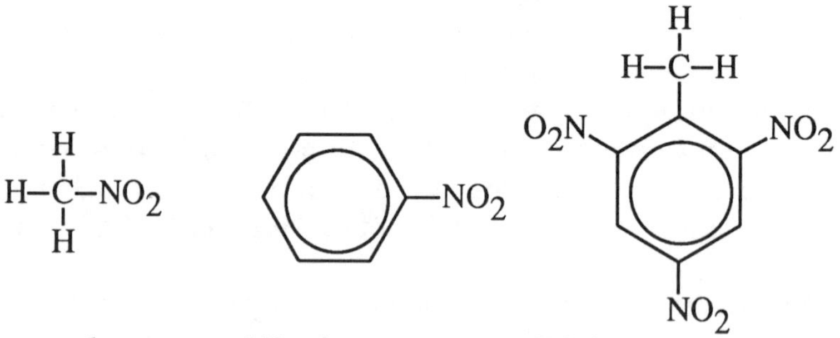

```
                                                 H
                                                 |
                                               H—C—H
                                                 |
     H                                  O2N    /   \    NO2
     |                    ____                |  O  |
 H—C—NO2                 ( O )—NO2             \   /
     |                                           |
     H                                          NO2

 Nitromethane         Nitrobenzene        Trinitrotoluene (TNT)
```

Figure 10.5. Some major nitro compounds.

The lightest of the nitro compounds is **nitromethane**, an oily liquid (mp -29°C, bp 101°C). It has a toxicity rating of 3. Symptoms of poisoning include anorexia, diarrhea, nausea, and vomiting. The organs that are most susceptible to damage from it are the kidneys and liver.

Nitrobenzene is a pale yellow oily liquid (mp 5.7°C, bp 211°C) with an odor of bitter almonds or shoe polish (mentioned in Section 3.9 as a symptom of nitrobenzene poisoning). It is produced mainly for the manufacture of aniline. It can enter the body through all routes and has a toxicity rating of 5. Its toxic action is much like that of aniline, including the conversion of hemoglobin to methemoglobin, which deprives tissue of oxygen. Cyanosis is a major symptom of nitrobenzene poisoning.

Trinitrotoluene (TNT) is a solid material widely used as a military explosive. It has a toxicity rating of 3–4. It can damage the cells of many kinds of tissue, including those of bone marrow, kidney, and liver. Extensive knowledge of the toxicity of TNT was obtained during the crash program to manufacture huge quantities of it during World War II. Toxic hepatitis developed in some workers under age 30 exposed to TNT systemically, whereas aplastic anemia was observed in some older victims of exposure. In the U.S. during World War II, 22 cases of fatal TNT poisoning were documented[5] (many more people were blown up during manufacture and handling).

Nitro Alcohols and Nitro Phenols

Nitro alcohols are nonaromatic compounds containing both –OH and –NO2 groups. A typical example of such a compound is **2-nitro-1-butanol**, shown below. These compounds are used in chemical synthesis to introduce nitro functional groups or (after reduction) amino groups onto molecules. They tend to have low volatilities and moderate toxicities. The aromatic nitrophenol, ***p*-nitrophenol**, is an industrially important compound with toxicological properties resembling those of phenol and nitrobenzene.

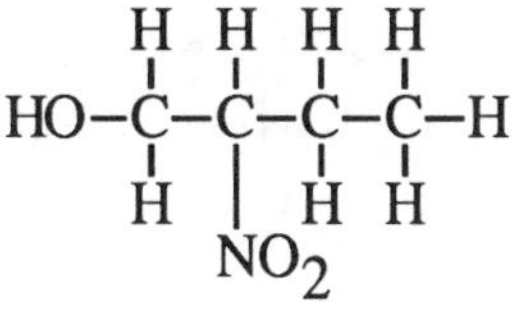

```
    H   H   H   H
    |   |   |   |
HO—C—C—C—C—H
    |   |   |   |
    H   |   H   H
       NO2
```

2–Nitro–1–butanol

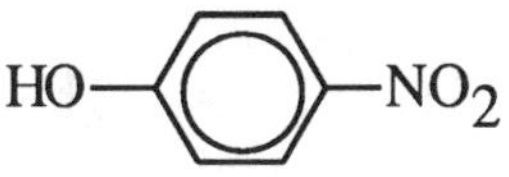

HO—⟨benzene ring⟩—NO_2

p– Nitrophenol

Dinoseb

Dinoseb is a nitrophenolic compound once widely used as a herbicide and plant dessicant that is noted for its toxic effects. The chemical name of this compound is 4,6-dinitro-2-*sec*-butylphenol, and its structure is shown below. Several derivatives of it have been marketed as herbicides.

Dinoseb has a toxicity rating of 5 and is strongly suspected of causing birth defects in the children of women exposed to it early in pregnancy, as well as sterility in exposed men. In October 1986 the

Environmental Protection Agency imposed an emergency ban on the use of the chemical, which was partially rescinded for the northwestern U. S. by court order early in 1987. In June, 1988, the E.P.A. allowed limited use of dinoseb through 1989, primarily in the northwestern U. S. for use on peas, chickpeas, lentils, and raspberry crops.[6]

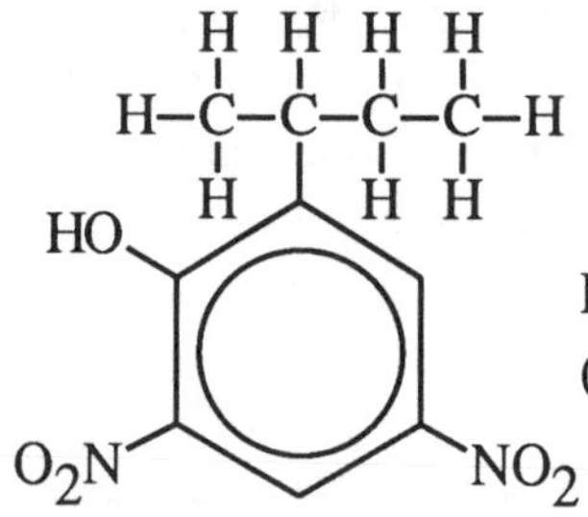

Dinoseb
(4,6–dinitro–2–*sec*–butylphenol)

10.7. NITROSAMINES

N-nitroso compounds, commonly called **nitrosamines**, are a class of compounds containing the N–N=O functional group. They are of particular toxicological significance because most that have been tested have been shown to be carcinogenic. The structures of some nitrosamines are shown in Figure 10.6.

Some nitrosamines have been used as solvents and as intermediates in chemical synthesis. They have been found in a variety of materials to which humans may be exposed, including beer, whiskey, and cutting oils used in machining.

By far the most significant toxicological effect of nitrosamines is their carcinogenicity, which may result from exposure to a single large dose or from chronic exposure to relatively small doses. Different nitrosamines cause cancer in different organs. The first nitrosamine extensively investigated for carcinogenicity was dimethylnitrosamine, once widely used as an industrial solvent. It was known to cause liver damage and jaundice in exposed workers[7] and studies starting in the 1950s subsequently revealed its carcinogenic nature. Dimethylnitrosamine was found to alkylate DNA, which is the mechanism of its carcinogenicity (the alkylation of DNA as a cause of cancer is noted in the discussion of biochemistry of carcinogesis in Section 4.6).

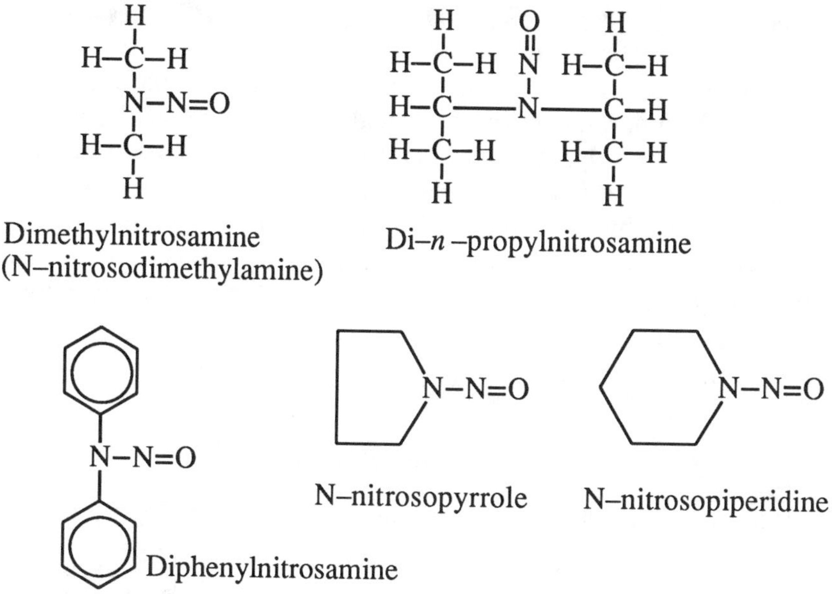

```
    H
    |
 H—C—H
    |
    N—N=O
    |
 H—C—H
    |
    H
```

Dimethylnitrosamine
(N–nitrosodimethylamine)

```
    H      O      H
    |      ‖      |
 H—C—H    N   H—C—H
    |      |      |
 H—C———N———C—H
    |             |
 H—C—H       H—C—H
    |             |
    H             H
```

Di–*n* –propylnitrosamine

N–N=O

Diphenylnitrosamine

N–N=O

N–nitrosopyrrole

N–N=O

N–nitrosopiperidine

Figure 10.6. Examples of some important nitrosamines.

The common means of synthesizing nitrosamines is the low-pH reaction of a secondary amine and nitrite as shown by the following example:

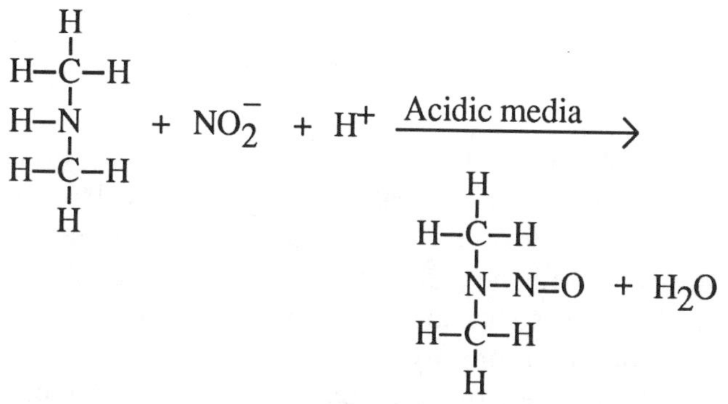

```
    H
    |
 H—C—H
    |
 H—N      + NO2⁻  + H⁺  —Acidic media→
    |
 H—C—H
    |
    H
                      H
                      |
                   H—C—H
                      |
                      N—N=O  + H2O        (10.3)
                      |
                   H—C—H
                      |
                      H
```

$$(CH_3)_2NH + NO_2^- + H^+ \xrightarrow{\text{Acidic media}} (CH_3)_2N\text{–}N{=}O + H_2O \quad (10.3)$$

The possibility of this kind of reaction occurring in vivo and producing nitrosamines in the acidic medium of the stomach is some cause for concern over nitrites in the diet. Because of this possibility nitrite levels have been reduced substantially in foods such as cured meats that formerly contained relatively high nitrite levels.

10.8. ISOCYANATES AND METHYL ISOCYANATE

Isocyanates are compounds with the general formula R–N=C=O. They have numerous uses in chemical synthesis, particularly in the manufacture of polymers with carefully tuned specialty properties. Methyl isocyanate is a raw material in the manufacture of carbaryl insecticide. Methyl isocyanate (like other isocyanates) can be synthesized by the reaction of a primary amine with phosgene in a moderately complex process represented by the following overall reaction:

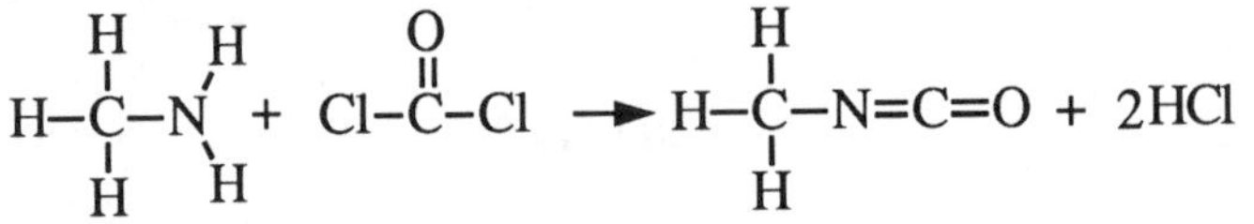

$$CH_3NH_2 + Cl-C(=O)-Cl \rightarrow CH_3-N=C=O + 2HCl \quad (10.4)$$

Methylamine Phosgene Methyl isocyanate

Structures of three significant isocyanates are given in Figure 10.7.

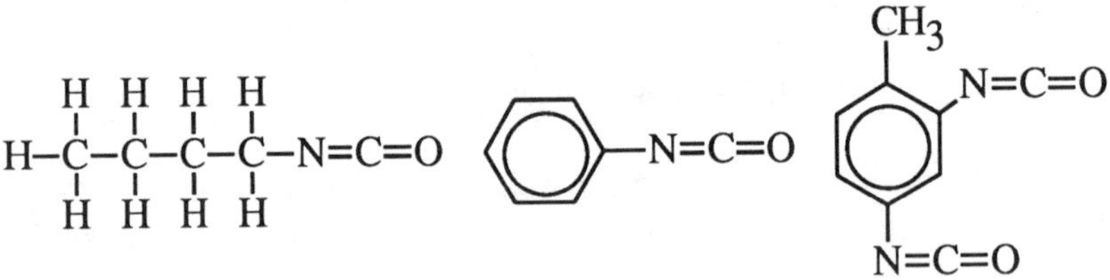

$CH_3CH_2CH_2CH_2-N=C=O$ — *n*–Butyl isocyanate

$C_6H_5-N=C=O$ — Phenyl isocyanate

$CH_3C_6H_3(N=C=O)_2$ — 2,4–Toluene diisocyanate

Figure 10.7. Example isocyanate compounds.

Both chemically and toxicologically, the most significant property of isocyanates is the high chemical reactivity of the isocyanate functional group. Industrially, the most significant such reaction is with alcohols to yield urethane (carbamate) compounds, as shown by Reaction 10.5. Multiple isocyanate and –OH groups in the reactant molecules enable formation of polymers. The chemical versatility of isocyanates and the usefulness of the products — such as polymers and pesticides — from which they are made have resulted in their widespread industrial production and consumption.

Methyl isocyanate was the toxic agent involved in the most catastrophic industrial accident of all time, which resulted in the release to the atmosphere of several tons of the compound in Bhopal,

India on December 2, 1984. More than 2,000 people were killed and about 100,000 suffered adverse effects.

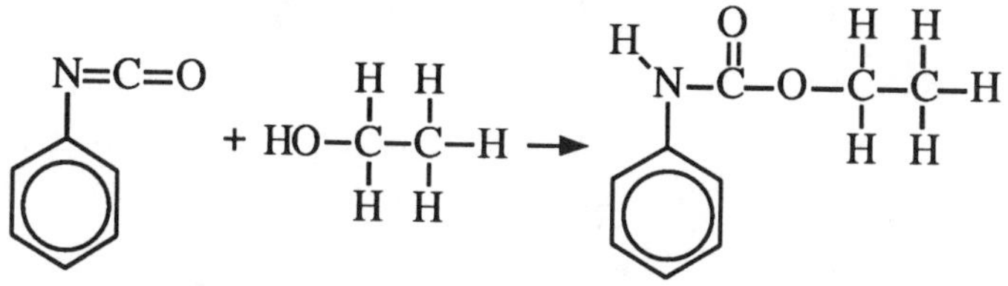

$$C_6H_5{-}N{=}C{=}O + HO{-}CH_2{-}CH_3 \rightarrow C_6H_5{-}NH{-}\overset{\displaystyle O}{\overset{\|}{C}}{-}O{-}CH_2{-}CH_3 \qquad (10.5)$$

Phenyl isocyanate — A carbamate or urethane compound

The major debilitating effects of methyl isocyanate on the Bhopal victims were on the lungs, with survivors suffering long-term shortness of breath and weakness from lung damage. However, victims also suffer symptoms of nausea and bodily pain and numerous toxic effects have been observed in the victims.[8] The tendency of the compound to function as a systemic poison appears to be the result of its ability to bind with small-molecule proteins and peptides. The most prominent among these is glutathione, a tripeptide described as a conjugating agent in Section 4.3. Isocyanate reacts reversibly with –SH groups on glutathione and the resulting complex can be transported in the body, enabling movement of isocyanate to various organs.

10.9. PESTICIDAL COMPOUNDS

A large number of organic compounds used as pesticides contain nitrogen. Space does not permit a detailed discussion of such compounds, but two general classes of them are cited here.

Carbamates

Pesticidal organic derivatives of carbamic acid, for which the formula is shown in Figure 10.8, are known collectively as **carbamates**. Carbamate pesticides have been widely used because some are more biodegradable than the formerly popular organochlorine insecticides and have lower dermal toxicities than most common organophosphate pesticides.

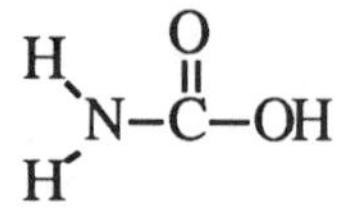

Carbamic acid

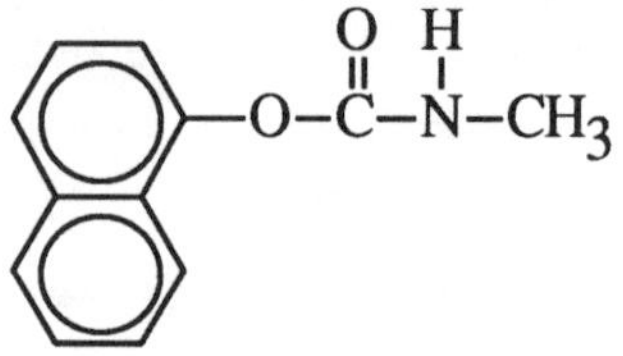

Carbaryl

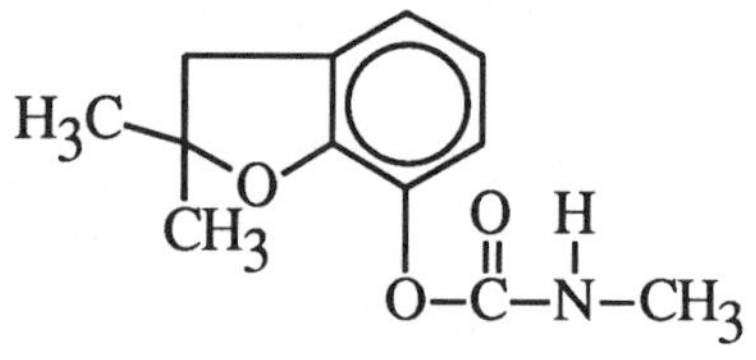

Carbofuran

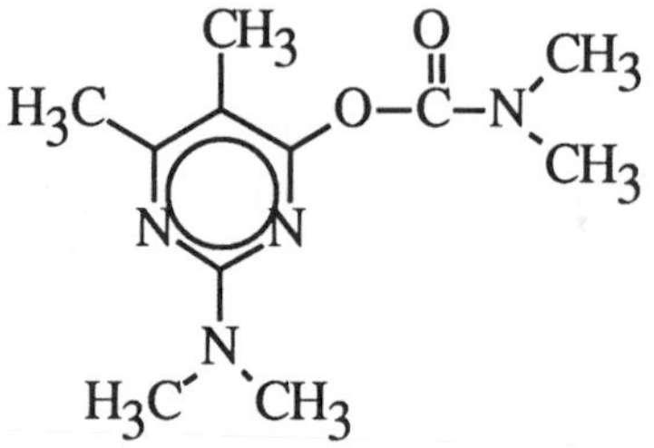

Pirimicarb

Figure 10.8. Carbamic acid and three insecticidal carbamates.

Carbaryl has been widely used as an insecticide on lawns or gardens. It has a low toxicity to mammals. **Carbofuran** has a high water solubility and acts as a plant systemic insecticide. It is taken up by the roots and leaves of plants so that insects feeding on the plant material are poisoned by the carbamate compound in it.

Pirimicarb has been widely used in agriculture as a systemic aphicide. Unlike many carbamates, it is rather persistent, with a strong tendency to bind to soil.[9]

The toxic effects of carbamates to animals are due to the fact that these compounds inhibit acetylcholinesterase. Unlike some of the organophosphate insecticides (see Chapter 13), they do so without the need for undergoing a prior biotransformation and are therefore classified as direct inhibitors. Their inhibition of acetylcholinesterase is relatively reversible. Loss of acetylcholinesterase inhibition activity may result from hydrolysis of the carbamate ester, which can occur metabolically. In general, carbamates have a wide range between a dose that causes onset of poisoning symptoms and a fatal dose (see discussion of dose-response in Sections 1.4 and 1.5). Although pirimicarb has a high systemic mammalian toxicity, its effects are mitigated by its low tendency to be absorbed through the skin.

Bipyridilium Compounds

As shown by the structures in Figure 10.9, a bipyridilium compound contains 2 pyridine rings per molecule. The two important pesticidal compounds of this type are the herbicides **diquat** and **paraquat**; other members of this class of herbicides include chlormequat, morfamquat, and difenzoquat. Applied directly to plant tissue, these compounds rapidly destroy plant cells and give the plant a frost-bitten appearance. However, they bind tenaciously to soil, especially the clay mineral fraction, which results in rapid loss of herbicidal activity so that sprayed fields can be planted within a day or two of herbicide application.

Paraquat, which was registered for use in 1965, is the most used of the bipyridilium herbicides. With a toxicity rating of 5, it is reputed to have "been responsible for hundreds of human deaths."[10] Exposure to fatal or dangerous levels of paraquat can occur by all pathways, including inhalation of spray, skin contact, ingestion, and even suicidal hypodermic injections. Despite these possibilities and its widespread application, paraquat is used safely without ill effects when proper procedures are followed.

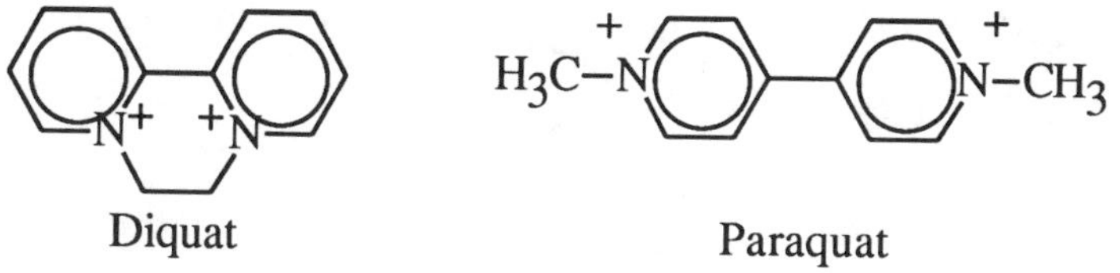

N^+ ^+N

Diquat

H_3C-N^+ N^+-CH_3

Paraquat

Figure 10.9. The two major bipyridilium herbicides (cation forms).

Because of its widespread use as a herbicide, the possibility exists of substantial paraquat contamination of food.[11] Drinking water contamination by paraquat has also been observed. The chronic effects of exposure to low levels of paraquat over extended periods of time are not well known. Acute exposure of animals to paraquat aerosols causes pulmonary fibrosis, and the lungs are affected even when exposure is through non-pulmonary routes. Paraquat affects enzyme activity. Acute exposure may cause variations in the levels of catecholamine, glucose, and insulin.

Although paraquat can be corrosive at the point of contact, it is a systemic poison that is devastating to a number of organs. The most prominent initial symptom of poisoning is vomiting, sometimes followed by diarrhea. Within a few days, dyspnea, cyanosis, and evidence of impairment of the kidneys, liver, and heart become obvious. In fatal cases, the lungs develop pulmonary fibrosis, often with pulmonary edema and hemorrhaging.

10.10. ALKALOIDS

Alkaloids are compounds of biosynthetic origin that contain nitrogen, usually in a heterocyclic ring. These compounds are produced by plants in which they are usually present as salts of organic acids.[12] They tend to be basic and to have a variety of physiological effects. Several alkaloids are shown earlier in the text. One of the more notorious ones, cocaine, was illustrated in Figure 3.6 and deadly poisonous strychnine is given in Figure 3.2. The structural formulas of three other alkaloids are given in Figure 10.10.

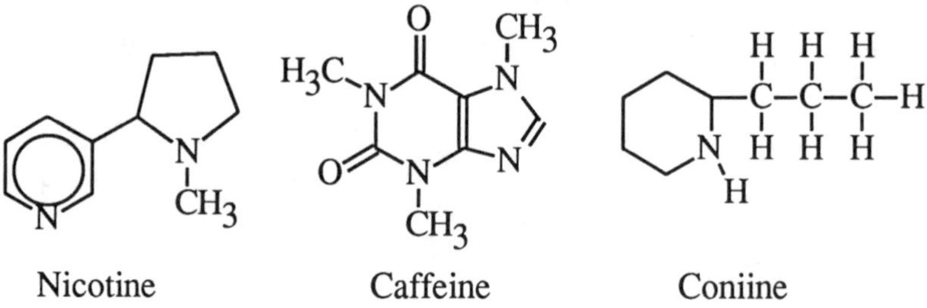

Nicotine Caffeine Coniine

Figure 10.10. Structural formulas of typical alkaloids.

Among the alkaloids are some well-known (and dangerous) compounds. Nicotine is an agent in tobacco that has been described as "one of the most toxic of all poisons and (it) acts with great rapidity."[13] In 1988 the U.S. Surgeon General declared nicotine to be an addictive substance. Coniine is the major toxic agent in poison hemlock (see Chapter 14). Alkaloidal strychnine is a powerful, fast-acting convulsant. Cocaine is currently the illicit drug of greatest concern. Quinine and sterioisomeric quinidine are alkaloids that are effective antimalarial agents. Like some other alkaloids, caffeine contains oxygen. It is a stimulant that can be fatal to humans in a dose of about 10 grams.

LITERATURE CITED

1. Schweizer, Albert E., "Amines," and Jean Northcott, "Amines, Aromatic," *Kirk-Othmer Concise Encyclopedia of Chemical Technology*, Wiley-Interscience, New York, 1985, pp. 82-86.

2. Williams, Phillip L., and James L. Burson, *Industrial Toxicology*, Van Nostrand Reinhold Co., New York, 1985.

3. Gosselin, Robert E., Roger P. Smith, and Harold C. Hodge, "Aniline," in *Clinical Toxicology of Commercial Products,* 5th ed., Williams and Wilkins, Baltimore/London, 1984, pp. III-32–III-36.

4. Smiley, Robert A., "Nitriles," *Kirk-Othmer Concise Encyclopedia of Chemical Technology*, Wiley-Interscience, New York, 1985, pp. 788-789.

5. Dreisbach, Robert E., and William O. Robertson, *Handbook of Poisoning*, 12th ed., Appleton and Lange, Norwalk, Connecticut, and Los Altos, California, 1987.

6. Shabecoff, Philip, "E.P.A. to Allow Limited Use of Chemical It Suspended," *New York Times,* June 11, 1988, p. 7.

7. Williams, Gary M., and John H. Weisburger, "Chemical Carcinogens," Chapter 5 in *Casarett and Doull's Toxicology*, 3rd ed., Curtis D. Klaassen, Mary O. Amdur, and John Doull, Eds., Macmillan Publishing Co., New York, 1986, p. 117.

8. Lepowski, Wil, "Methyl Isocyanate: Studies Point to Systemic Effects," *Chemical and Engineering News*, June 13, 1988, p. 6.

9. Sanchez-Camazano, M., and M. J. Sanchez-Martin, "Influence of Soil Characteristics on the Adsorption of Pirimicarb," *Environmental Toxicology and Chemistry*, **7**, 559-564 (1988).

10. Gosselin, Robert E., Roger P. Smith, and Harold C. Hodge, "Paraquat," in *Clinical Toxicology of Commercial Products,*

5th ed., Williams and Wilkins, Baltimore/London, 1984, pp. III-328–III-336.

11. Van Emon, Jeanette, James Seiber, and Bruce Hammock, "Application of an Enzyme-Linked Immunosorbent Assay (ELISA) to Determine Paraquat Residues in Milk, Beef, and Potatoes," *Bull. Environ. Contam. Toxicol.*, **39**, 490-497 (1987).

12. Cordell, Godfrey A., "Alkaloids," in *Kirk-Othmer Concise Encyclopedia of Chemical Technology*, Wiley-Interscience, New York, 1985, pp. 63-67.

13. Gosselin, Robert E., Roger P. Smith, and Harold C. Hodge, "Nicotine," in *Clinical Toxicology of Commercial Products,* 5th ed., Williams and Wilkins, Baltimore/London, 1984, pp. III-311–III-314.

11

Organohalide Compounds

1.1. INTRODUCTION

Organohalide compounds are halogen-substituted hydrocarbons with a wide range of physical and chemical properties produced in large quantities as solvents, heat transfer fluids, chemical intermediates, and for other applications.[1] They may be saturated (alkyl halides), unsaturated (alkenyl halides), or aromatic (aryl halides). The major means of synthesizing organohalide compounds were shown by examples in Chapter 8 and include substitution halogenation, addition halogenation, and hydrohalogenation reactions, illustrated in Equations 8.2, 8.7, and 8.9, respectively. Most organohalide compounds are chlorides (chlorocarbons and chlorohydrocarbons), but they also include compounds of fluorine, bromine, and iodine, as well as mixed halides, such as the chlorofluorocarbons.

The chemical reactivities of organohalide compounds vary over a wide range. The alkyl halides are generally low in reactivity, but may undergo pyrolysis in flames to liberate noxious products, such as HCl gas. Alkenyl halides may be oxidized, which in some cases produces highly toxic phosgene, as shown by the following example:

$$\underset{\text{Trichloroethylene}}{ClHC{=}CCl_2} + O_2 \rightarrow HCl + \underset{\text{Phosgene}}{Cl{-}\overset{\overset{\displaystyle O}{\|}}{C}{-}Cl} + CO \qquad (11.1)$$

The toxicities of organohalide compounds vary widely. For example, dichlorodifluoromethane ("Freon–12") is generally regarded as having a low toxicity, except for narcotic effects and the possibility of asphyxiation at high concentration. Vinyl chloride (see Section 11.3), however, is a known human carcinogen. The polychlorinated biphenyls are highly resistant to biodegradation and are extremely persistent in the environment.

11.2. ALKYL HALIDES

Alkyl halides are compounds in which halogen atoms are substituted for hydrogen on an alkyl group. The structural formulas of some typical alkyl halides are given in Figure 11.1. Most of the commercially important alkyl halides are derivatives of alkanes of low molecular mass.

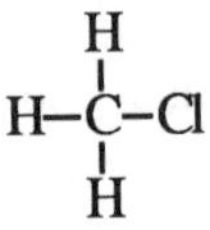

```
   H              H               Cl
   |              |               |
H–C–Cl        Cl–C–Cl         Cl–C–Cl
   |              |               |
   H              H               Cl
```

Chloromethane (fp -98°C, bp -24°C)

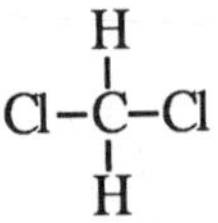

Dichloromethane (methylene chloride, fp -97°C, bp 40°C)

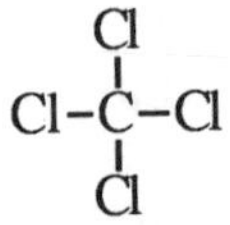

Carbon tetrachloride (fp -23°C, bp 77°C)

```
   F              H  H            Cl H
   |              |  |            |  |
Cl–C–Cl        H–C–C–Cl       Cl–C–C–H
   |              |  |            |  |
   F              H  H            Cl H
```

Dichlorodifluoromethane (“Freon-12,” fp -158°C, bp -29°C)

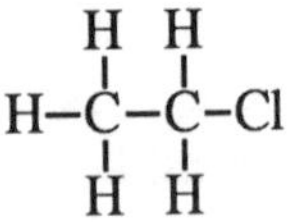

Chloroethane (ethylene chloride, fp -139°C, bp 12°C)

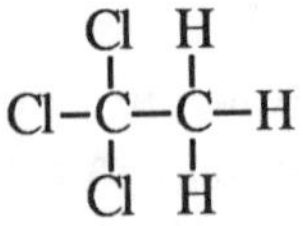

1,1,1–Trichloroethane (methyl chloroform, fp -33°C, bp 74°C)

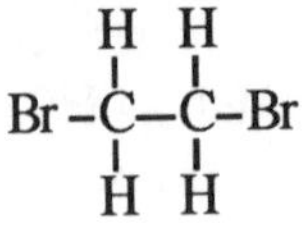

```
    H  H
    |  |
Br -C–C-Br
    |  |
    H  H
```

1,2–Dibromoethane (ethylene dibromide, fp 9.3°C, bp 131°C)

Figure 11.1. Some typical low-molecular-mass alkyl halides.

A brief discussion of the uses of the compounds listed in Figure 11.1 will provide an idea of the versatility of the alkyl halides. Volatile chloromethane (methyl chloride) was once widely used as a refrigerant fluid and aerosol propellant; most of it now is consumed in the manufacture of silicones. Dichloromethane is a volatile liquid with excellent solvent properties for nonpolar organic solutes. It has been applied as a solvent for the decaffeination of coffee and in paint strippers, as a blowing agent in urethane polymer manufacture, and to depress vapor pressure in aerosol formulations. Once commonly sold as a solvent and stain remover, carbon tetrachloride is now largely restricted to uses as a chemical intermediate under controlled conditions, primarily to manufacture chlorofluorocarbon refrigerant fluid compounds (for example, dichlorodifluoromethane). Chloroethane is an intermediate in the manufacture of tetraethyllead and is an ethylating agent in chemical synthesis. One of the more common industrial chlorinated solvents is 1,1,1-trichloroethane. Insecticidal 1,2-dibromoethane has been used in large quantities to fumigate soil, grain, and fruit and as a lead scavenger in leaded gasoline. It is an effective solvent for resins, gums, and waxes and serves as a chemical intermediate in the syntheses of some pharmaceutical compounds and dyes.

Toxicities of Alkyl Halides

The toxicities of alkyl halides vary a great deal with the compound. Although some of these compounds were considered to be almost completely safe in the past, there is a marked tendency to regard each with more caution as additional health and animal toxicity study data become available. Perhaps the most universal toxic effect of alkyl halides is depression of the central nervous system. Chloroform ($CHCl_3$) was the first widely used general anesthetic, although many surgical patients were accidentally killed by it.

Carbon Tetrachloride and Lipid Peroxidation

Of all the alkyl halides, carbon tetrachloride has the most notorious record of human toxicity.[2] For many years it was widely used in consumer products as a degreasing solvent, in home fire extinguishers, and for other applications. However, numerous toxic effects, including some fatalities, were observed and in 1970 the U. S. Food and Drug Administration (FDA) banned the sale of carbon tetrachloride and formulations containing it for home use.

Carbon tetrachloride is toxic through both inhalation and ingestion. Toxic symptoms from inhalation tend to be associated with the nervous system, whereas those from ingestion often involve the gastrointestinal tract and liver. Both the liver and kidneys may be substantially damaged by carbon tetrachloride.

The biochemical mechanism of carbon tetrachloride toxicity has been investigated in detail.[3] The cytochrome P-450-dependent monooxygenase system acts on CCl_4 in the liver to produce the $Cl_3C\cdot$ free radical:

$$CCl_4 \rightarrow Cl_3C\cdot \qquad (11.2)$$

This product can combine with molecular oxygen to yield highly reactive $Cl_3COO\cdot$ radical:

$$CCl_4 + O_2 \rightarrow Cl_3C{-}OO \qquad (11.3)$$

These radical species, along with others produced from their subsequent reactions, can react with biomolecules, such as proteins and DNA. The most damaging such reaction is **lipid peroxidation**, a process that involves the attack of chemically active species on unsaturated lipid molecules, followed by oxidation of the lipids through a free radical mechanism. It occurs in the liver and is a major mode of action of some hepatotoxicants, which can result in substantial cellular damage.[4] The mechanism of lipid peroxidation is not known with certainty. It is known that the methylene hydrogens attached to doubly bonded carbon atoms in lipid molecules are subject to abstraction by free radicals as shown by the following:

$$\text{HC=CH (Lipid Molecule, L)} + Cl_3C\cdot \rightarrow \text{HC=C}\cdot \text{ (Lipid radical, L}\cdot\text{)} + Cl_3C{-}H \qquad (11.4)$$

Reaction of the lipid radical with molecular oxygen yields peroxy radical species:

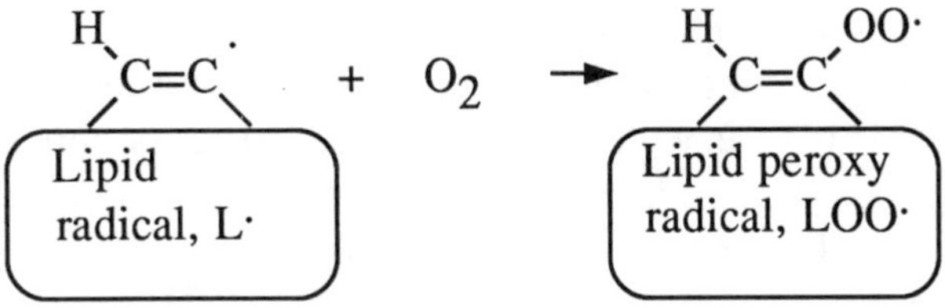

H C=C· (Lipid radical, L·) + O_2 → H, OO· C=C (Lipid peroxy radical, LOO·) (11.5)

This species can initiate chain reaction sequences with other molecules as follows:

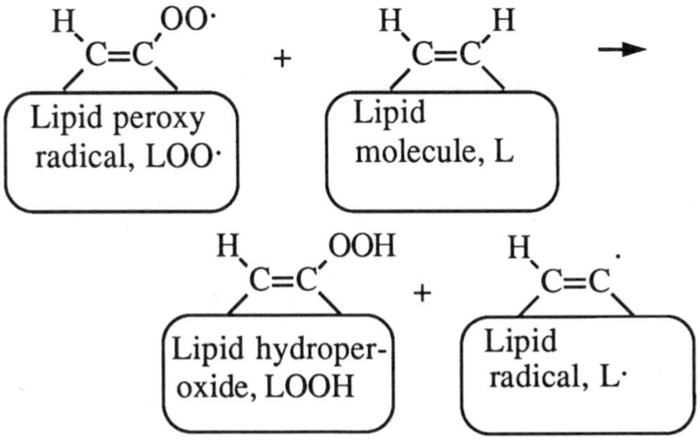

H, OO· C=C (Lipid peroxy radical, LOO·) + H, H C=C (Lipid molecule, L) →

H, OOH C=C (Lipid hydroperoxide, LOOH) + H C=C· (Lipid radical, L·) (11.6)

Once initiated, chain reactions such as these continue and cause massive alteration of the lipid molecules. The LOOH molecules are unstable and decompose to yield additional free radicals. The process terminates when free radical species combine with each other to form stable species.

Toxicities of Other Alkyl Halides

Dichloromethane has long been regarded as one of the least acutely toxic alkyl halides. More volatile than most commonly used solvents, this compound has been used in large quantities as a degreasing solvent, paint remover, aerosol propellant additive, and grain fumigant. As a result, human exposure has been relatively high. In 1987, however, the U. S. Occupational Safety and Health Administration (OSHA) considered a move to substantially lower the permissible human exposure limit on evidence that dichloromethane is a probable human carcinogen.[5]

Generally considered to be among the least toxic of the alkyl halides, 1,1,1-trichloroethane is widely used. Because so many people may be exposed to this compound, any toxic effects are of concern. A much more toxic alkyl halide is 1,2-dibromoethane. It is a severe irritant, damages the lungs when inhaled in high concentrations, and is a potential human carcinogen, so its use has been severely curtailed.

11.3. ALKENYL HALIDES

The **alkenyl** or **olefinic organohalides** contain at least one halogen atom and at least one carbon–carbon double bond. The most significant of these are the lighter chlorinated compounds, such as those illustrated in Figure 11.2.

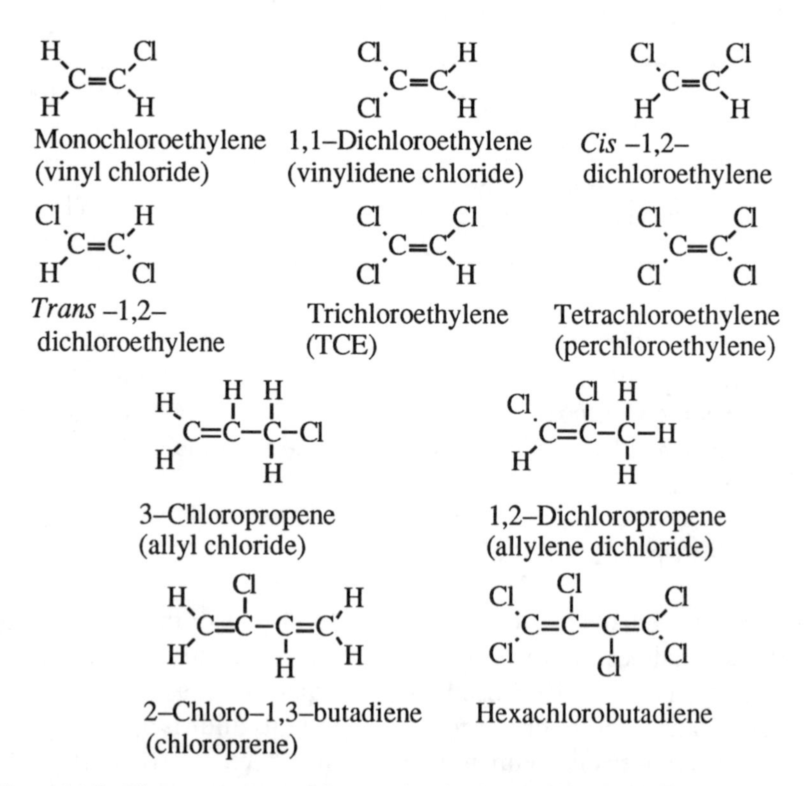

Figure 11.2. The more common low-molecular-mass alkenyl chlorides.

Uses of Alkenyl Halides

The alkenyl halides are used for numerous purposes. Some of the more important applications are discussed here.

Vinyl chloride is consumed in large quantities to manufacture polyvinylchloride plastic, a major polymer in pipe, hose, wrapping, and other products. Vinyl chloride is a highly flammable volatile gas with a sweet, somewhat pleasant odor.

As shown in Figure 11.2, there are three possible dichloroethylene compounds, all clear, colorless liquids. Vinylidene chloride forms a copolymer with vinyl chloride used in some kinds of coating materials. The geometrically isomeric 1,2-dichloroethylenes are used as organic synthesis intermediates and as solvents.

Trichloroethylene is an excellent solvent for organic substances and has some other properties that are favorable for a solvent. It is a clear, colorless, nonflammable, volatile liquid. It is an excellent degreasing and drycleaning solvent and has been used as a household solvent and for food extraction (for example, in decaffeination of coffee).

Tetrachloroethylene is a colorless, non-flammable liquid with properties similar to those of trichloroethylene. Its major use is for drycleaning, and it has some applications for degreasing metals.

The two chlorinated propene compounds shown are colorless liquids with pungent, irritating odors. Allyl chloride is an intermediate in the manufacture of allyl alcohol and other allyl compounds, including pharmaceuticals, insecticides, and thermosetting varnish and plastic resins. Dichloropropene compounds can be used as soil fumigants, as well as solvents for oil, fat, drycleaning, and metal degreasing.

Chloroprene is produced in large quantities for the manufacture of neoprene rubber. It is a colorless liquid with an ethereal odor. Hexachlorobutadiene is a colorless liquid with an odor somewhat like that of turpentine. It is used as a solvent for higher hydrocarbons and elastomers, as a hydraulic fluid, in transformers, and for heat transfer.

Toxic Effects of Alkenyl Halides

Because of their widespread use and disposal in the environment, the toxicities of the alkenyl halides are of considerable concern. They exhibit a wide range of acute and chronic toxic effects.

Many workers have been exposed to vinyl chloride because of its use in polyvinylchloride plastic manufacture. The central nervous system, respiratory system, liver, and blood and lymph systems are all affected by exposure to vinyl chloride. Among the symptoms of poisoning are fatigue, weakness, and abdominal pain. Cyanosis may also occur. Vinyl chloride was abandoned as an anesthetic when it was found to induce cardiac arrhythmias.

The most notable effect of vinyl chloride is its carcinogenicity. It causes a rare angiosarcoma of the liver in chronically exposed individuals, observed particularly in those who cleaned autoclaves in the polyvinylchloride fabrication industry. The carcinogenicity of vinyl chloride results from its metabolic oxidation to chloroethylene oxide by the action of the cytochrome P-450 monooxygenase enzyme system in the liver[3] as follows:

$$\mathrm{H_2C{=}CHCl} + \{\mathrm{O}\} \rightarrow \mathrm{H{-}\underset{H}{C}{-}\!\overset{O}{\frown}\!{-}\underset{H}{C}{-}Cl} \xrightarrow{\text{Rearrangement}} \mathrm{Cl{-}\underset{H}{\overset{H}{C}}{-}\overset{O}{\overset{\|}{C}}{-}H} \qquad (11.7)$$

Chloroacetaldehyde

The epoxide has a strong tendency to covalently bond to protein, DNA, and RNA, and it rearranges to chloroacetaldehyde, a known mutagen. Therefore, vinyl chloride produces two potentially carcinogenic metabolites. Both of these products can undergo conjugation with glutathione to yield products that are eliminated from the body.

Based upon animal studies and its structural similarity to vinyl chloride, 1,1-dichloroethylene is a suspect human carcinogen. Although both 1,2-dichloroethylene isomers have relatively low toxicities, their modes of action are different. The *cis* isomer is an irritant and narcotic, whereas the *trans* isomer affects both the central nervous system and the gastrointestinal tract, causing weakness, tremors, cramps, and nausea.

Trichloroethylene has caused liver carcinoma in experimental animals and is a suspect human carcinogen. Numerous body organs are affected by it. As with other organohalide solvents, skin dermatitis can result from dissolution of skin lipids by trichloroethylene. Exposure to it can affect the central nervous and respiratory systems, liver, kidneys, and heart. Symptoms of exposure include disturbed vision, headaches, nausea, cardiac arrhythmias, and burning/tingling sensations in the nerves (paresthesia).

Tetrachloroethylene damages the liver, kidneys, and central nervous system. Because of its hepatotoxicity and experimental evidence of carcinogenicity in mice, it is a suspect human carcinogen.

The chlorinated propenes are obnoxious compounds. Unlike other compounds discussed so far in this section, their pungent odors and irritating effects lead to an avoidance response in exposed subjects. They are irritants to the eyes, skin, and respiratory tract. Contact with the skin can result in rashes, blisters, and burns. Chronic exposure to allyl chloride is manifested by aching muscles and bones; it damages the liver, lungs, and kidney and causes pulmonary edema.

Chloroprene is an eye and respiratory system irritant. It causes dermatitis to the skin and alopecia, a condition characterized by hair loss in the affected skin area. Affected individuals are often nervous and irritable.

Ingestion and inhalation of hexachlorobutadiene inhibits cells in the liver and kidney. Animal tests have shown both acute and chronic toxicities. The compound is a suspect human carcinogen.

Hexachlorocyclopentadiene

As shown by the structure below, hexachlorocyclopentadiene is a cyclic alkenyl halide with two double bonds:

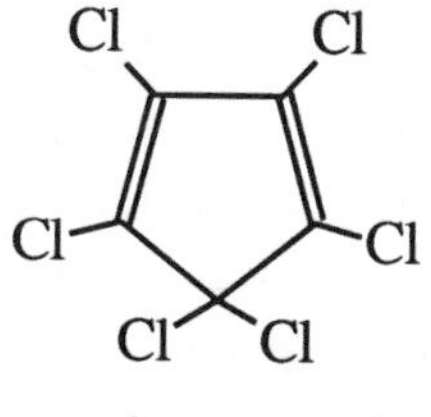

Cl Cl

Cl Cl

Cl Cl

Hexachlorocyclopentadiene

It was once an important industrial chemical used directly as an agricultural fumigant and as an intermediate in the manufacture of insecticides. Hexachlorocyclopentadiene and still bottoms from its manufacture are found in hazardous waste chemical sites, and large quantities were disposed of at the Love Canal site. The pure compound is a light yellow liquid (fp 11°C, bp 239°C, density 1.7 g/cm^3) with a pungent, somewhat musty odor. With two double bonds, it is a very reactive compound and readily undergoes substitution and addition reactions. Its photolytic degradation yields water-soluble products.

Hexachlorocyclopentadiene is considered to be very toxic, with a toxicity rating of 4. Its fumes are strongly lachrimating, and it is a skin, eye, and mucuous membrane irritant. In experimental animals it has been found to damage most major organs, including the kidney, heart, brain, adrenal glands, and liver.

11.4. ARYL HALIDES

Figure 11.3 gives the structural formulas of some of the more important aryl halides. These compounds are made by the substitution chlorination of aromatic hydrocarbons as shown, for example, in Figure 8.3 for the synthesis of a polychlorinated biphenyl.

Properties and Uses of Aryl Halides

Aryl halides have many uses, which have resulted in substantial human exposure and environmental contamination. Some of their major applications are summarized here.

Monochlorobenzene is a flammable clear liquid (fp -45°C, bp 132°C) used as a solvent, as a solvent carrier for methylene diisocyanate, as a pesticide, as a heat transfer fluid, and in the manufacture of aniline, nitrobenzene, and phenol. The 1,2- isomer of dichlorobenzene (*ortho*-dichlorobenzene) has been used as a solvent for degreasing hides and wool and as a raw material for dye manufacture; the 1,4- isomer (*para*-dichlorobenzene) is also used in dye manufacture and as a moth repellant and germicide; and all three isomers have been used as fumigants and insecticides. The 1,2- and 1,3- (*meta*) isomers are liquids under ambient conditions, whereas the 1,4- isomer is a white sublimable solid. Used as a solvent, lubricant, dielectric fluid, chemical intermediate, and formerly as a termiticide, 1,2,4-trichlorobenzene is a liquid (fp 17°C, bp 213°C).

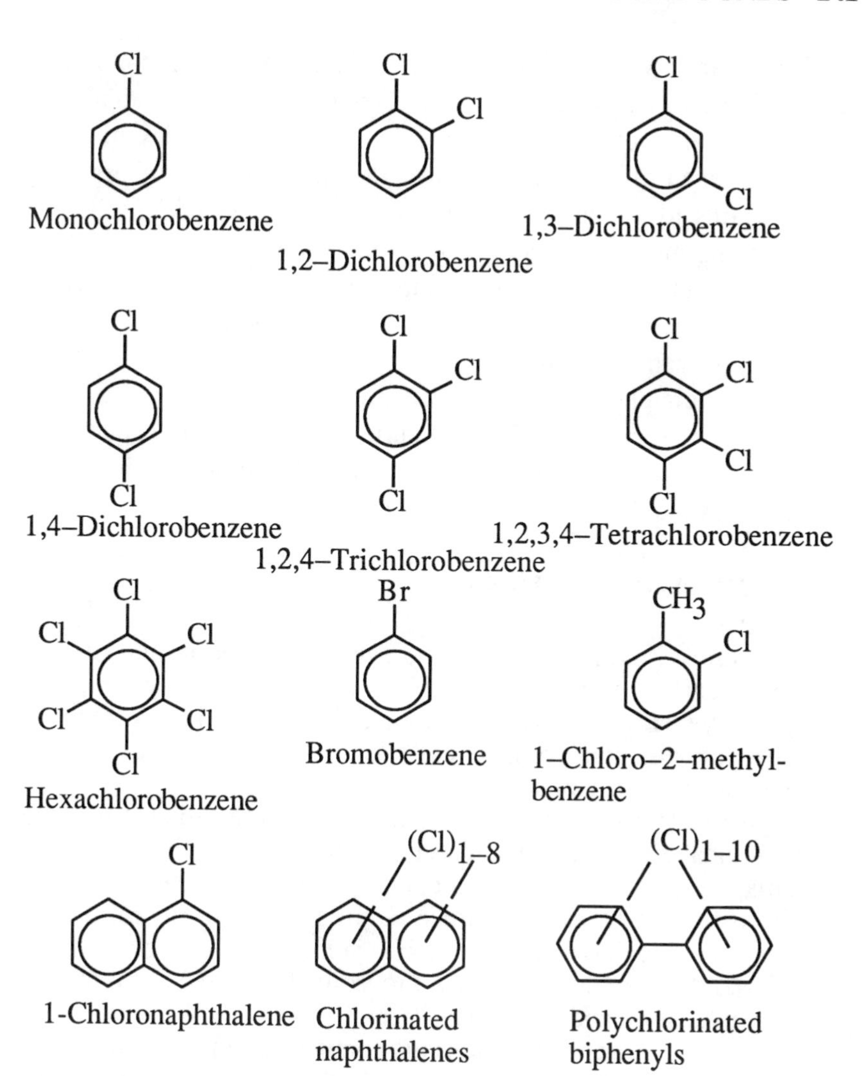

Figure 11.3. Some of the more important aryl halides.

Hexachlorobenzene (perchlorobenzene), is a high-melting-point solid consisting of white needles and is used as a seed fungicide, wood preservative, and intermediate for organic synthesis. Bromobenzene (fp -31°C, bp 156°C) serves as a solvent and motor oil additive, as well as an intermediate for organic synthesis. Most 1-chloro-2-methylbenzene is consumed in the manufacture of 1-chlorobenzo-trifluoride.

There are two major classes of halogenated aryl compounds containing two benzene rings. One class is based upon naphthalene and the other upon biphenyl, as shown by the examples in Figure 11.3. For each class of compounds, the individual members range from liquids to solids, depending upon the degree of chlorination. These compounds are manufactured by chlorination of the parent compounds and have been sold as mixtures with varying degrees of chlorine content. The desirable properties of the chlorinated naphthalenes, polychlorinated biphenyls, and polybrominated biphenyls, including their physical and chemical stabilities, have led to many uses, such as for heat transfer, hydraulic fluids, dielectrics, and flame retardants. However, for environmental reasons these uses have been severely curtailed.

Toxic Effects of Aryl Halides

Exposure to monochlorobenzene usually occurs by inhalation or skin contact. It is an irritant and affects the respiratory system, liver, skin, and eyes. Ingestion of this compound has caused incoordination, pallor, cyanosis, and eventual collapse, effects similar to those of aniline poisoning (see Section 10.3).

Exposure to the dichlorobenzenes is also most likely to occur through inhalation or contact. These compounds are irritants and tend to damage the same organs as monochlorobenzene. The 1,4- isomer has been known to cause profuse rhinitis (running nose), nausea, weight loss associated with anorexia, jaundice, and liver cirrhosis. The di- and tetrachlorobenzenes are considered to be moderately toxic by inhalation and ingestion.

Hexachlorobenzene is a notorious compound in the annals of toxicology because of a massive poisoning incident involving 3,000 people in Turkey during the period 1955–1959.[6] The victims ate seed wheat that had been treated with 10% hexachlorobenzene to deter fungal growth. As a consequence, they developed **porphyria cutanea tarda**, a condition in which the skin becomes blistered, fragile, photosensitive, and subject to excessive hair growth. In addition to the skin damage, the victims' eyes were damaged in severe cases and many suffered weight loss associated with anorexia. Wasting of skeletal muscles was also observed. The possibility exists that many of these effects were due to the presence of manufacturing by-product impurity polychlorinated dibenzodioxins (see Section 11.6).

Bromobenzene can enter the body through the respiratory tract, gastrointestinal tract, or skin. Little information is available regarding its human toxicity. It has been shown to damage the livers of rats used in animal tests.

Wide variations have been reported in the toxicities of the chlorinated naphthalenes, raising the possibility that some of the effects observed were due to impurities introduced during manufacture. Humans exposed to the more highly halogenated fractions by inhaling the vapors have developed chloracne rash and have suffered from debilitating liver necrosis. In the 1940s and early 1950s several hundred thousand cattle died from polychlorinated naphthalene-contaminated feed.

Polychlorinated biphenyls (PCBs) are of concern because of their widespread environmental occurrence and extreme persistence.[7] Their polybrominated biphenyl analogs (PBBs) were the cause of massive livestock poisoning in Michigan in 1973 because of the addition of PBB flame retardant to livestock feed during its formulation.

11.5. ORGANOHALIDE INSECTICIDES

Organohalide compounds were the first of the widely used synthetic organic pesticides. In this section organohalide insecticides are discussed and in Section 11.6 other pesticides of the organohalide chemical type are covered.

Figure 11.4 shows the structural formulas of some of the more common organohalide insecticides. Most of the insecticidal organohalide compounds contain chlorine as the only halogen. Ethylene dibromide and dichlorobromopropane are insecticidal, but are more properly classified as fumigants and nematocides.

As seen from the structural formulas in Figure 11.4, the organochlorine insecticides are of intermediate molecular mass and contain at least one aromatic or nonaromatic ring. They can be placed in four major chemical classes. The first of these consists of the chloroethylene derivatives, of which DDT and methoxychlor are the major examples. The second group is composed of chlorinated cyclodiene compounds, including aldrin, dieldrin, and heptachlor. The most highly chlorinated members of this class, such as chlordecone, are manufactured from hexachlorocyclopentadiene (see Section 11.3).

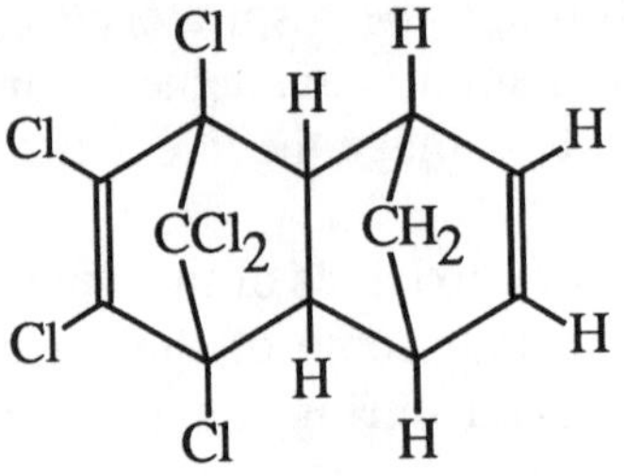

Aldrin

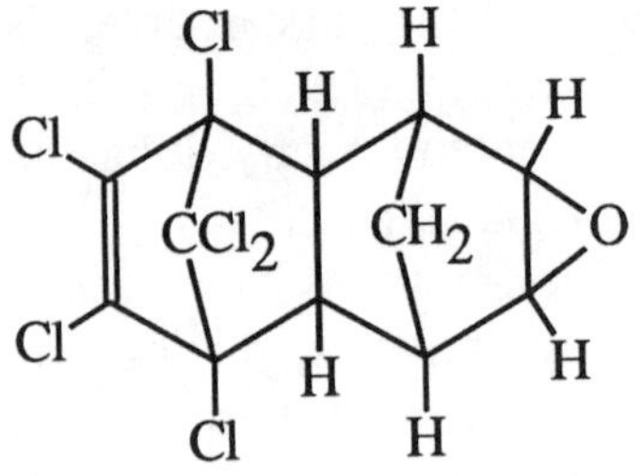

Dieldrin

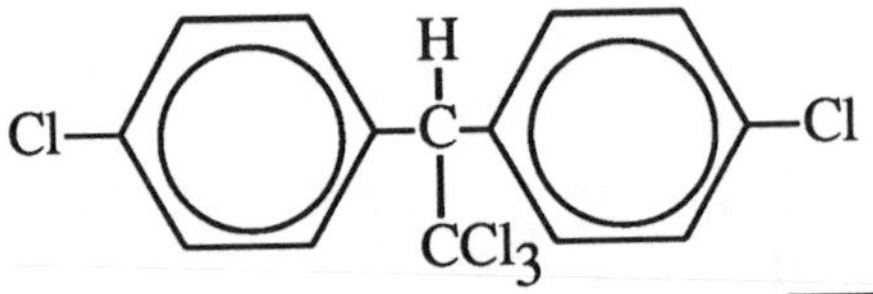

DDT

Methoxychlor

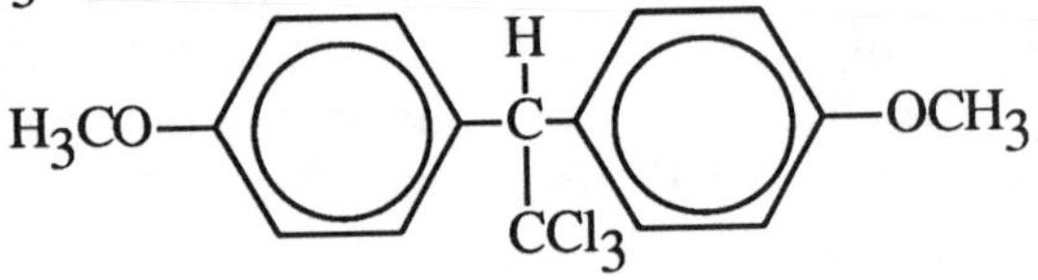

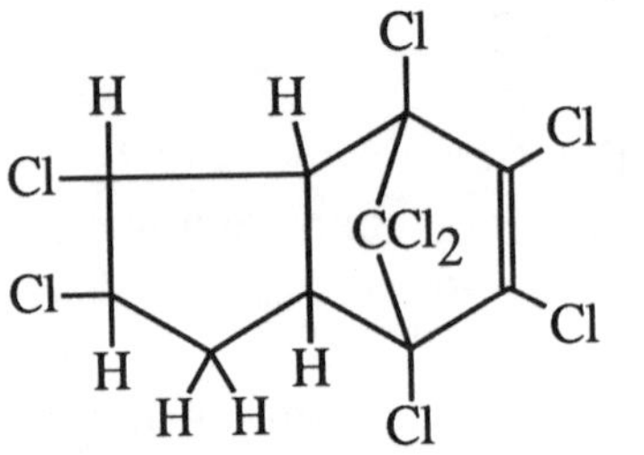

Chlordane

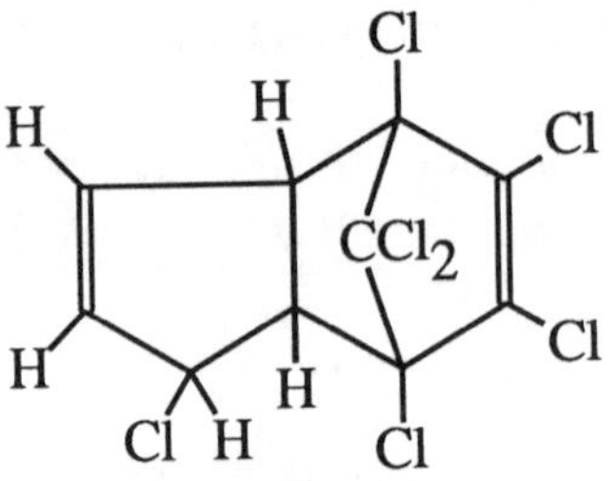

Heptachlor

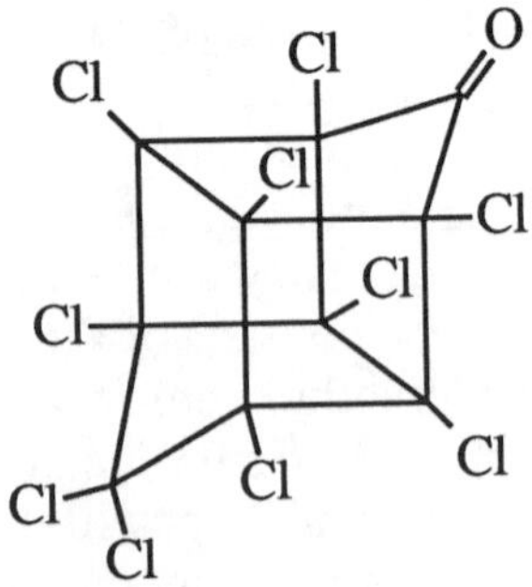

Chlordecone (Kepone); Mirex is structurally similar with 2 Cl's in place of =O.

Figure 11.4. Some typical organohalide insecticides.

The benzene hexachloride sterioisomers make up a third class of organochlorine insecticides and the final group, known collectively as toxaphene, constitutes a fourth.

Toxicities of Organohalide Insecticides

Organohalide insecticides exhibit a wide range of toxic effects and varying degrees of toxicity. Many of these compounds are neuropoisons and their most prominent acute effects are upon the central nervous system, with symptoms of CNS poisoning including tremor, irregular jerking of the eyes, changes in personality, and loss of memory.[8] Some of the toxic effects of specific organohalide insecticides and classes of these compounds are discussed below.

Despite its role in the establishment of the modern environmental movement through Rachel Carson's classic book *Silent Spring*, the acute toxicity of DDT to humans is very low. It was applied directly to people on a large scale during World War II for the control of typhus and malaria. Symptoms of acute DDT poisoning are much the same as those described previously for organohalide insecticides in general and are for the most part neurotoxic in nature. In the environment, DDT undergoes bioaccumulation in the food chain, with animals at the top of the chain most affected. The most vulnerable of these are predator birds that produce thin-shelled, readily broken eggs from ingestion of DDT through the food chain. The other major insecticidal chloroethane-based compound, methoxychlor, is a generally more biodegradable, less toxic compound than DDT, and has been used as a substitute for it.

The toxicities of the chlorinated cyclodiene insecticides, including aldrin, dieldrin, endrin, chlordane, heptachlor, endosulfan, and isodrin, are relatively high and similar to each other. They appear to act on the brain, releasing betaine esters and causing headaches, dizziness, nausea, vomiting, jerking muscles, and convulsions. Some members of this group are teratogenic or toxic to fetuses. In test animals, dieldrin, chlordane, and heptachlor have caused liver cancer. For several years the use of aldrin, dieldrin, and heptachlor has been prohibited in the U. S., and chlordane was restricted to underground applications for termite control. In 1987, even this use of chlordane was discontinued.

A significant number of human exposures to the insecticides

derived from hexachlorocyclopentadiene (Mirex and Kepone) have occurred. Use of these environmentally damaging compounds has been restricted in the U. S. to eradication of fire ants in the southeastern states. The manufacture of Kepone in Hopewell, Virginia, during the 1970s resulted in the discharge of about 53,000 kg of this compound to the James River through the city sewage system. Toxic effects of Kepone include central nervous system symptoms (irritability, tremor, hallucinations), adverse effects on sperm, and damage to the nerves and muscles. The compound causes liver cancer in rodents and is teratogenic in test animals. Studies of exposed workers have shown that Kepone absorbed by the liver is excreted through the bile, then reabsorbed from the gastrointestinal tract, thereby participating in the enterohepatic circulation system as illustrated in Figure 3.4.

Hexachlorocyclohexane

Hexachlorocyclohexane, often confusingly called benzene hexachloride (BHC), consists of several sterioisomers with different orientations of H and Cl atoms. The gamma isomer is shown in Figure 11.5. It is an effective insecticide, constituting at least 99% of the commercial insecticide **lindane**.

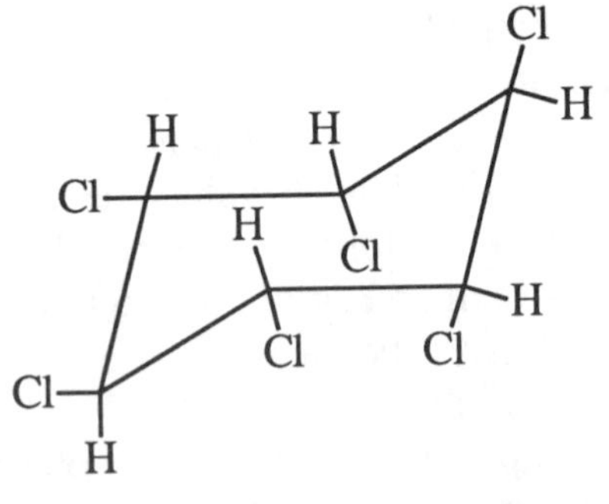

Figure 11.5. The gamma isomer of hexachlorocyclohexane (lindane).

The toxic effects of lindane are very similar to those of DDT. Degeneration of kidney tubules, liver damage associated with fatty tissue, and hystoplastic anemia have been observed in individuals poisoned by lindane.

Toxaphene

Toxaphene is insecticidal chlorinated camphene and consists of a mixture of more than 170 compounds containing 10 C atoms and 6–10 Cl atoms per molecule and often represented by the empirical formula $C_{10}H_{10}Cl_8$. The structural formula of one of the molecules contained in toxaphene, 8-octachlorobornane, is given below. Toxaphene was once the most widely used insecticide in the U. S. with annual consumption of about 40 million kg.

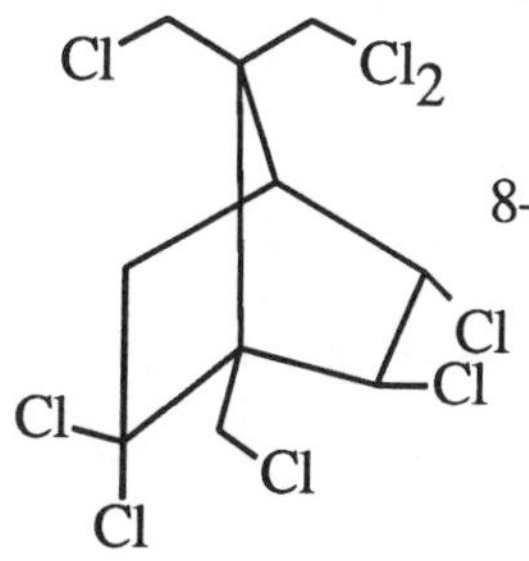

8–Octachlorobornane

The many compounds found in formulations of toxaphene vary widely in their toxicities. One of the most toxic of these compounds is 8-octachlorobornane, shown above. Toxaphene produces convulsions of an epileptic type in exposed mammals.

11.6. NONINSECTICIDAL ORGANOHALIDE PESTICIDES

The best known noninsecticidal organohalide pesticides are the **chlorophenoxy** compounds. These consist of 2,4-dichlorophenoxyacetic acid (2,4-D), 2,4,5-trichlorophenoxyacetic acid (2,4,5-T or Agent Orange), and a closely related compound, Silvex. These compounds, their esters, and their salts have been used as ingredients of a large number of herbicide formulations. Formulations of 2,4,5-T have been made notorious largely by a manufacturing by-product, 2,3,7,8-tetrachloro-*p*-dioxin (TCDD, commonly known as "dioxin"). The structural formulas of these compounds are shown in Figure 11.6.

Toxic Effects of Chlorophenoxy Herbicides

The oral toxicity rating of 2,4-dichlorophenoxyacetic acid is 4, although the toxicities of its commercially marketed ester and salt forms are thought to be somewhat lower. Large doses have been shown to cause nerve damage, such as peripheral neuropathy, as well as convulsions and even brain damage. A National Cancer Institute study of Kansas farmers who had handled 2,4-D extensively has shown an occurrence of non-Hodgkins lymphoma 6 to 8 times that of comparable unexposed populations.[9] The toxicity of Silvex appears to be somewhat less than that of 2,4-D and to a large extent it is excreted unchanged in the urine.

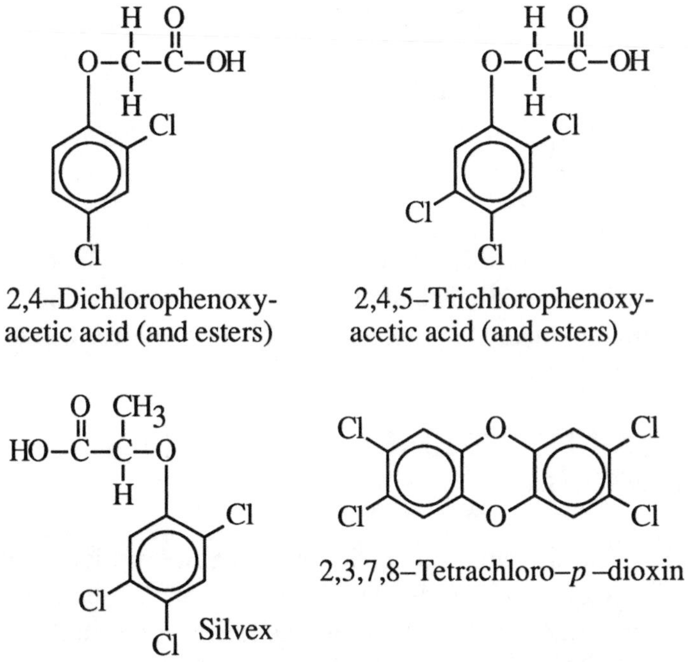

2,4–Dichlorophenoxyacetic acid (and esters)

2,4,5–Trichlorophenoxyacetic acid (and esters)

Silvex

2,3,7,8–Tetrachloro–*p*–dioxin

Figure 11.6. Herbicidal chlorophenoxy compounds and TCDD manufacturing by-product.

Although the toxic effects of 2,4,5-T may even be somewhat less than those of 2,4-D, observations of 2,4,5-T toxicity have been complicated by the presence of manufacturing by-product TCDD. Experimental animals dosed with 2,4,5-T have exhibited mild spasticity. Some fatal poisonings of sheep have been caused by 2,4,5-T herbicide. Autopsied carcasses revealed nephritis, hepatitis, and enteritis. Humans absorb 2,4,5-T rapidly and excrete it largely unchanged through the urine.

Toxicity of TCDD

TCDD belongs to the class of compounds called **polychlorinated dibenzodioxins**, which have the same basic structure as TCDD, but different numbers and arrangements of chlorine atoms on the ring structure. These compounds exhibit varying degrees of toxicity. Classified as a supertoxic compound, TCDD is unquestionably extremely toxic to some animals. Its acute LD_{50} to male guinea pigs is only 0.6 μg/kg of body mass. Because of its production as a manufacturing by-product of some commercial products such as 2,4,5-T, possible emission from municipal incineration, and widespread distribution in the environment from improper waste disposal (for example, the infamous "dioxin" spread from waste oil at Times Beach, Missouri) or discharge from industrial accidents (Seveso, Italy), TCDD has become a notorious environmental pollutant. However, the degree and nature of its toxicity to humans are both rather uncertain. It is known to cause a human skin condition called chloracne.

In a December, 1987, action, the U.S. Environmental Protection Agency altered its estimate of the potential cancer-causing risk of TCDD to 1/16 its previous value. This was a tentative action subject to additional review and scrutiny.

The most massive industrial accident in which TCDD was released occurred at the Givaudan-La Roche Icmesa manufacturing plant near Seveso, Italy, in 1976. Several tens of thousands of people were exposed when a cloud of chemical emissions spread over an approximately 3-square-mile area. In a 1988 report covering 15,291 children born in the area within 6 years after the release, researchers at Catholic University (Rome) and the Italian Birth Defects Monitoring Program revealed that there was not evidence of any increase in malformations among the children relative to a control group.[10] These findings applied as well to children born within 9 months of the chemical release whose mothers were in the exposure area. Of further interest was the finding that no major malformation was found in children from the most highly contaminated area.

Alachlor

Widely marketed as Monsanto's Lasso® herbicide, Alachlor (Figure 11.7) has become a widespread contaminant of groundwater in some corn- and soybean-producing areas. It seems to be efficiently absorbed through the skin. Allergic skin reactions and skin and eye irritation have been reported in exposed individuals. The EPA has estimated a lifetime cancer risk of 1 in 100,000 from drinking water containing 2 ppb of alachlor,[11] although this risk estimate has been disputed by manufacturers.

Chlorinated Phenols

The chlorinated phenols, particularly **pentachlorophenol** (Figure 11.7) and the trichlorophenol isomers, have been widely used as wood

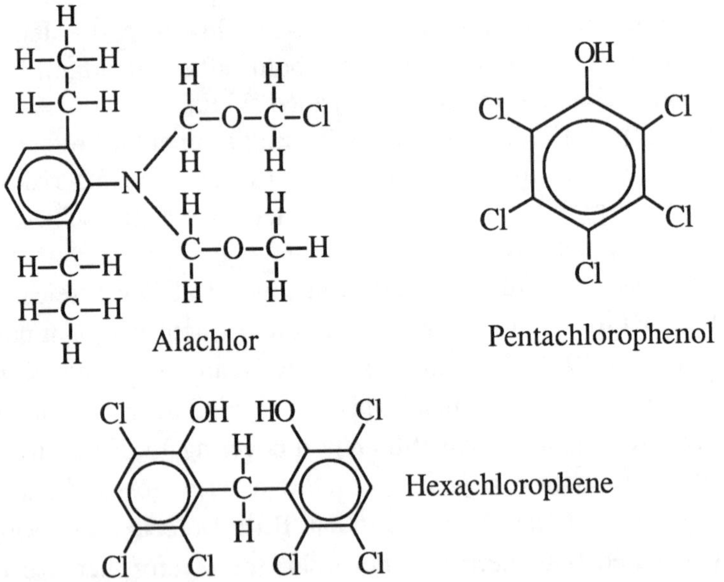

Alachlor

Pentachlorophenol

Hexachlorophene

Figure 11.7. Structural formulas of alachlor, pentachlorophenol, and hexachlorophene.

preservatives. Applied to wood, these compounds prevent wood rot through their fungicidal action and prevent termite infestation because of their insecticidal properties. Both cause liver malfunction and dermatitis. Contaminant polychlorinated dibenzodioxins may be responsible for some of the observed effects.

Hexachlorophene

Hexachlorophene (Figure 11.7) has been used as an agricultural fungicide and bactericide, largely in the production of vegetables and cotton. It is most noted for its use as an antibacterial agent in personal care products, now discontinued because of toxic effects and possible TCDD contamination.

LITERATURE CITED

1. Archer, Wesley L., "Chlorocarbons and Chlorohydrocarbons," in *Kirk-Othmer Concise Encyclopedia of Chemical Technology*, Wiley-Interscience, New York, 1985, p. 261.

2. Gosselin, Robert E., Roger P. Smith, and Harold C. Hodge, "Carbon Tetrachloride," in *Clinical Toxicology of Commercial Products,* 5th ed., Williams and Wilkins, Baltimore/London, 1984, pp. III-101–III-107.

3. Hodgson, Ernest, and Patricia E. Levi, *Modern Toxicology*, Elsevier, New York, 1987.

4. Hodgson, Ernest, and Frank E. Guthrie, Eds., *Introduction to Biochemical Toxicology*, Elsevier, New York, 1980.

5. "Dropping Methylene Chloride," *Industrial Chemist*, July, 1987, p. 20.

6. Gosselin, Robert E., Roger P. Smith, and Harold C. Hodge, "Hexachlorobenzene," in *Clinical Toxicology of Commercial Products,* 5th ed., Williams and Wilkins, Baltimore/London, 1984, pp. II-170–II-171.

7. Safe, S., Ed., *Polychlorinated Biphenyls (PCBs): Mammalian and Environmental Toxicology*, Springer-Verlag, New York, 1987.

8. Stopford, Woodhall, "The Toxic Responses of Pesticides," Chapter 11 in *Industrial Toxicology*, Phillip L. Williams and James L. Burson, Eds., Van Nostrand Reinhold Co., New York, 1985, pp. 211–229.

9. Silberner, J., "Common Herbicide Linked to Cancer," *Science News*, **130**(11), 167–174 (1986).

10. "Dioxin is Found Not to Increase Birth Defects," *New York Times*, March 18, 1988, p. 12.

11. "EPA Proposal on Alachlor Nears," *Science*, **233**, 1143–1144 (1986).

12

Sulfur-Containing Organic Compounds

12.1. INTRODUCTION

Sulfur is directly below oxygen in the periodic table. The sulfur atom has 6 valence electrons as shown in Figure 12.1 and its electron configuration is {Ne}$3s^2 3p^4$. Because of their very similar valence shell electron configurations, oxygen and sulfur behave somewhat alike chemically. However, unlike oxygen, the sulfur atom has three underlying 3*d* orbitals, and its valence shell can be expanded to more than 8 electrons. This makes sulfur's chemical behavior more diverse than that of oxygen. For example, sulfur has several common oxidation states, including -2, +4, and +6, whereas most chemically combined oxygen is in the -2 oxidation state.

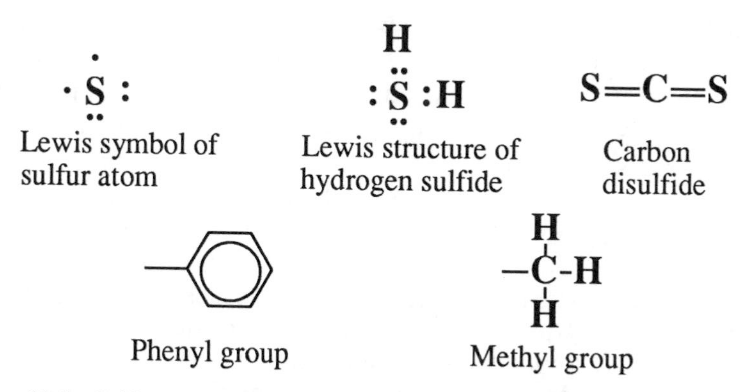

Figure 12.1. Sulfur atom, compounds, and substituent groups.

Classes of Organosulfur Compounds

The hydride of sulfur is H_2S (Figure 12.1), a highly toxic gas discussed in Section 7.11. Substitution of alkyl or aryl hydrocarbon groups such as phenyl and methyl (Figure 12.1) for H on hydrogen sulfide leads to a number of different organosulfur compounds. These include thiols (R–SH) and thioethers (R–S–R). Because of the availability of 3*d* orbitals, sulfur that is bonded to hydrocarbon moieties can also be bonded to oxygen, adding to the variety of organosulfur compounds that can exist.

Despite the high toxicity of H_2S, not all organosulfur compounds are particularly toxic. Many of the compounds have strong, offensive odors that warn of their presence, which reduces their hazard.

12.2. THIOLS, SULFIDES, AND DISULFIDES

Substitution of alkyl and aryl groups for H on H_2S yields **thiols** and **sulfides** (thioethers). Structural formulas of examples of these compounds are shown in Figure 12.2.

Thiols

Thiols are also known as mercaptans. The lighter alkyl thiols, such as methanethiol, are fairly common air pollutants with odors that may be described as "ultragarlic." Inhalation of even very low concentrations of the alkyl thiols in air can be very nauseating and result in headaches. Exposure to higher levels can cause increased pulse rate, cold hands and feet, and cyanosis. With extreme cases, unconsciousness, coma, and death may occur. The biochemical action of alkyl thiols likely is similar to that of H_2S and they are precursors to cytochrome oxidase poisons.

Gaseous methanethiol and volatile liquid ethanethiol (bp 35°C) are intermediates in pesticide synthesis and odorants placed in lines and tanks containing natural gas, propane, and butane to warn of leaks. Information about their toxicities to humans is lacking, although these compounds and 1-propanethiol should be considered dangerously toxic, especially by inhalation. Both 1- and 2-butanethiol are associated with skunk odor. Also known as amyl mercaptan, 1-pentanethiol (bp 124°C) is an allergen and weak sensitizer which causes contact dermatitis.

A typical alkenyl mercaptan is 2-propene-1-thiol, also known as allyl mercaptan. It is a volatile liquid (bp 68°C) with a strong garlic odor. It has a high toxicity and is strongly irritating to mucous membranes when inhaled or ingested.

Alpha-toluenethiol, also called benzyl mercaptan (bp 195°C) is very toxic orally. It is an experimental carcinogen.

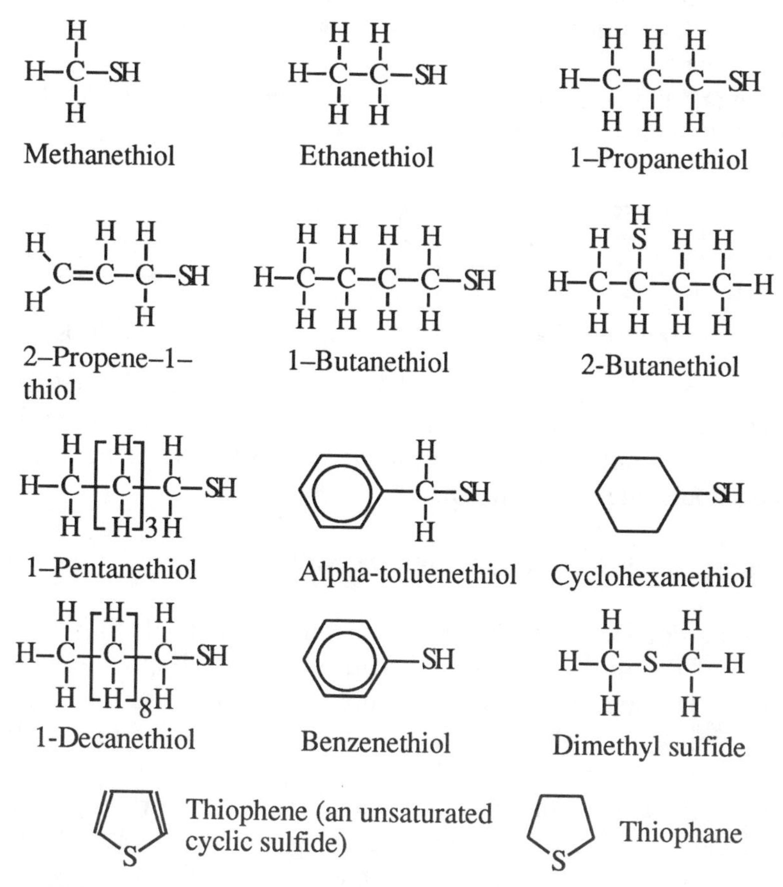

Figure 12.2. Common low-molecular-mass thiols and sulfides. All are liquids at room temperature, except for methanethiol, which boils at 5.9°C.

The simplest of the aryl thiols is benzenethiol (phenyl mercaptan), bp 168°C. It has a severely "repulsive" odor. Inhalation causes headache and dizziness and skin exposure results in severe contact dermatitis.

Sulfides and Disulfides

Dimethyl sulfide is an alkyl sulfide or thioether. It is a volatile liquid (bp 38°C) that is moderately toxic by ingestion. Thiophene is the most common cyclic sulfide. It is a heat-stable liquid (bp 84°C) with a solvent action much like that of benzene. It is used in the manufacture of pharmaceuticals and dyes, as well as resins that also contain phenol or formaldehyde. Its saturated analog is tetrahydrothiophene, or thiophane.

The organic disulfides contain the –SS– functional group as shown in the following two examples:

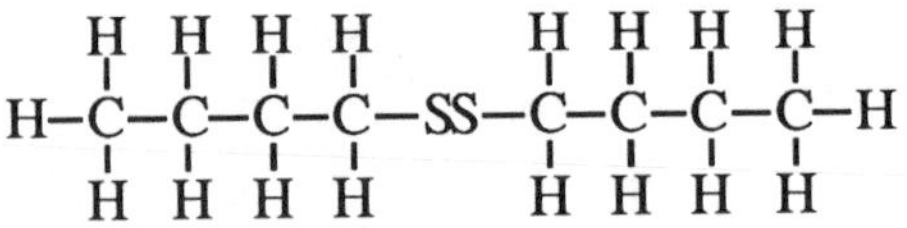

```
    H  H  H  H      H  H  H  H
    |  |  |  |      |  |  |  |
H–C–C–C–C–SS–C–C–C–C–H          ⟨⟩–SS–⟨⟩
    |  |  |  |      |  |  |  |
    H  H  H  H      H  H  H  H
```

n–Butyldisulfide 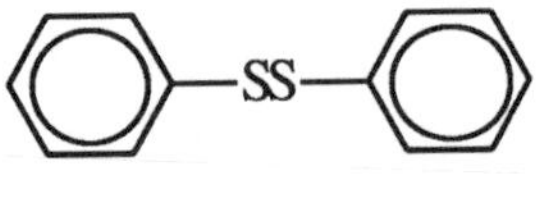   Diphenyldisulfide

These compounds may act as allergens that produce dermatitis in contact with skin. Not much information is available regarding their toxicities to humans, although animal studies suggest several toxic effects, including homolytic anemia.

Carbon Disulfide and Carbon Oxysulfide

Of the simple organic compounds of sulfur, **carbon disulfide** (CS_2) is one of the most significant because of its widespread use and toxicity. Despite its name, it does not contain the disulfide (–SS–) group, but has 2 sulfur atoms each separately bonded to a carbon atom. This compound is a volatile colorless liquid (mp -111°C, bp 46°C). Unlike most organosulfur compounds, it is virtually free of odor. Although its uses are declining, it has numerous applications in chemical synthesis, as a solvent to break down cellulose in viscose rayon manufacture, and in the manufacture of cellophane. It has also been used as an insecticide and fumigant.

Acute doses of carbon disulfide inhaled at 100–1000 ppm irritate mucous membranes and affect the central nervous system, usually causing excitation as a first noticeable effect, followed by restlessness, depression, and stupor. It is a much stronger anesthetic than chloroform (Section 11.2), causing unconsciousness and even death in cases of high exposure. Symptoms experienced during recovery from severe acute carbon disulfide poisoning resemble those that occur following intoxication from ingestion of ethanol in alcoholic beverages.

Chronic carbon disulfide poisoning by absorption through the skin or respiratory tract involves the central and peripheral nervous systems and may cause anemia. Symptoms include indistinct vision, neuritis, and a bizarre sensation of "crawling" on the skin.[1] Psychopathological symptoms may be varied and severe, including excitation, depression, irritability, and general loss of mental capabilities to the point of insanity. Parkinsonian paralysis may result from chronic carbon disulfide poisoning.

Replacement of one of the S atoms on carbon disulfide with an O atom yields **carbon oxysulfide** (COS) a volatile liquid boiling at 50°C. It can decompose to liberate toxic hydrogen sulfide. Carbon oxysulfide vapor is a toxic irritant. At high concentrations this compound has a strong narcotic effect.

12.3. ORGANOSULFUR COMPOUNDS CONTAINING NITROGEN

Several important classes of organosulfur compounds contain nitrogen. These compounds are discussed in this section.

Thiourea Compounds

Thiourea is the sulfur analog of urea. Substitution of hydrocarbon moieties on the N atoms yields various organic derivatives of thiourea as illustrated in Figure 12.3. Thiourea has been used as a rodenticide. It has a moderate to high toxicity to humans, affecting bone marrow and causing anemia. It has been shown to cause liver and thyroid cancers in experimental animals.

Phenylthiourea is likewise a rodenticide. Its toxicity is highly selective to rodents relative to humans, although it probably is very toxic to some other animals. The compound is metabolized extensively, and some of the sulfur is excreted as sulfate in urine.

Commonly called ANTU, **1-naphthylthiourea** is a virtually tasteless rodenticide that has a very high rodent:human toxicity ratio. The lethal dose to monkeys is about 4,000 mg/kg. One suicidal adult male human ingested about 80 g of 30% ANTU rat poison along with a considerable amount of alcohol. He vomited soon after ingestion and survived without significant ill effects.[2] Dogs, however, are quite susceptible to ANTU poisoning.

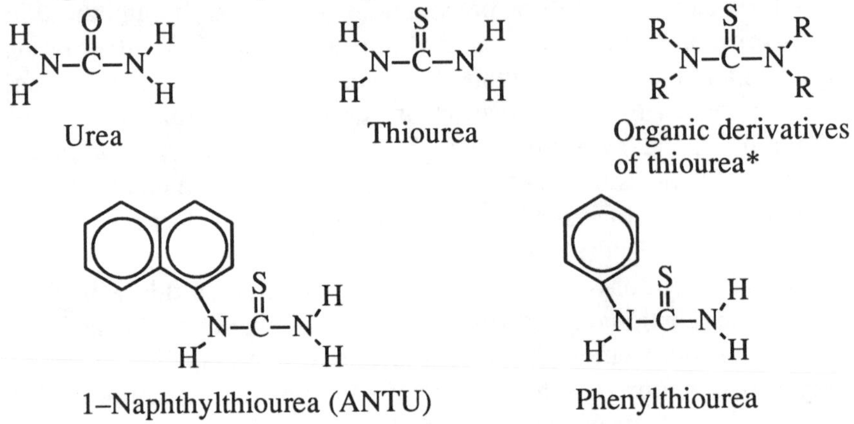

$$\mathrm{H_2N{-}C({=}O){-}NH_2}$$

Urea

$$\mathrm{H_2N{-}C({=}S){-}NH_2}$$

Thiourea

$$\mathrm{R_2N{-}C({=}S){-}NR_2}$$

Organic derivatives of thiourea*

$$\mathrm{C_{10}H_7{-}NH{-}C({=}S){-}NH_2}$$

1–Naphthylthiourea (ANTU)

$$\mathrm{C_6H_5{-}NH{-}C({=}S){-}NH_2}$$

Phenylthiourea

Figure 12.3. Structural formulas of urea, thiourea, and organic derivatives of thiourea.

* At least one R group is an alkyl, alkenyl, or aryl substituent.

Thiocyanates

Organic **thiocyanates** are derivatives of thiocyanic acid (HSCN) in which the H is replaced by hydrocarbon moieties, such as the methyl group. Dating from the 1930s and regarded as the first synthetic organic insecticides, these compounds kill insects upon contact. Because of their volatilities, the lower-molecular-mass methyl, ethyl, and isopropyl thiocyanates are effective fumigants for insect control. Insecticidal lauryl thiocyanate (below) is not volatile and is used in sprays in petroleum-based solvents and in dusting powders.

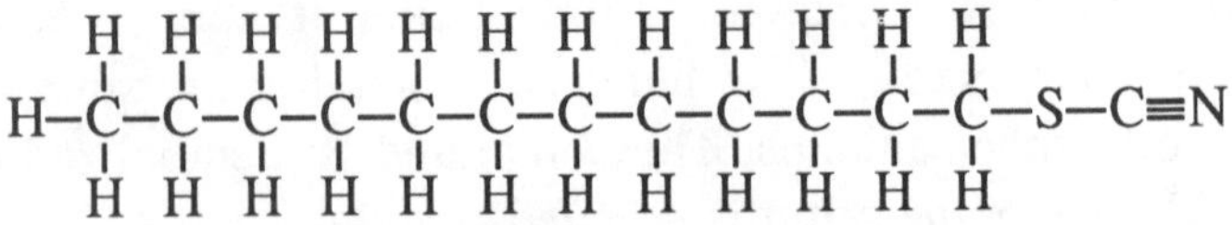

$$\mathrm{H{-}(CH_2)_{12}{-}S{-}C{\equiv}N}$$

Lauryl thiocyanate

The toxicities of the thiocyanates vary widely by compound and route of administration. Some metabolic processes liberate HCN from thiocyanates, which can result in death. Therefore, methyl, ethyl, and isopropyl thiocyanates should be regarded as rapid-acting, potent poisons.

The **isothiocyanate** group is illustrated in the structure of methyl-isothiocyanate below:

```
    H
    |
H—C—N=C=S   Methylisothiocyanate
    |
    H
```

Other compounds in this class include ethyl, allyl, and phenyl isothiocyanates. Methylisocyanate, also known as methyl mustard oil, and its ethyl analog have been developed as military poisons. Both are powerful irritants to eyes, skin, and respiratory tract. When decomposed by heat, these compounds emit sulfur oxides and hydrogen cyanide.

Disulfiram

Disulfiram is a sulfur- and nitrogen-containing compound with several industrial uses, including applications as a rubber accelerator and vulcanizer, fungicide, and seed disinfectant. It is most commonly known as **antabuse**, a therapeutic agent for the treatment of alcohol abuse which causes nausea, vomiting, and other adverse effects when ethanol is ingested.

```
    H  H                          H  H
    |  |                          |  |
H—C—C                            C—C—H
    |  | \                      / |  |
    H  H  \     S       S      /  H  H
           \    ||      ||    /
            N—C—SS—C—N
           /                  \
    H  H  /                    \  H  H
    |  | /                      \ |  |
H—C—C                            C—C—H
    |  |                          |  |
    H  H   Disulfiram (antabuse)  H  H
```

Disulfiram (antabuse)

Cyclic Sulfur/Nitrogen Organic Compounds

The structural formulas of several cyclic compounds containing both nitrogen and sulfur are shown in Figure 12.4. Basic to the structures of these compounds is the simple ring structure of **thiazole**.

It is a colorless liquid (bp 117°C). One of its major uses has been for the manufacture of sulfathiazole, one of the oldest of the sulfonamide class of antibacterial drugs. The use of sulfathiazole is now confined to the practice of veterinary medicine because of its serious side effects.[3]

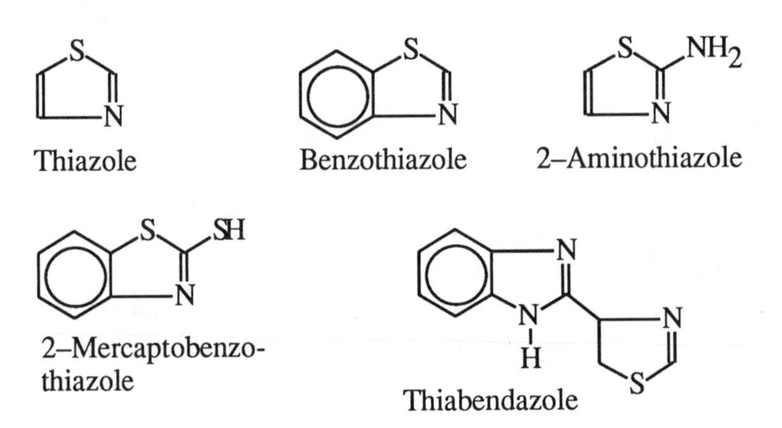

Figure 12.4. Cyclic compounds containing nitrogen and sulfur.

Several derivatives of thiazole have commercial uses. One of these is **2-aminothiazole**, which has shown a high toxicity to experimental animals. **Benzothiazole** is another related compound used in organic synthesis. Its human toxicity is not known, although it has a high toxicity to mice. **Thiabendazole** (2-(4'-thiazoyl)benzimidazole) is a systemic fungicide that can be carried through a plant and onto plant leaves.[4] Rats and dogs tolerate a relatively high dose of this chemical, although it tends to make the latter vomit. Acting as an adjuvant (Section 1.11), 2-mercaptobenzothiazole is mixed with dithio-carbamate fungicides to enhance their effects.

Dithiocarbamates

Dithiocarbamate fungicides consist of metal salts of **dimethylthio-carbamate** and **ethylenebisdithiocarbamate** anions as shown in Figure 12.5. These fungicides are named in accordance with the metal ion present. For example, the manganese salt of dimethyldithiocarbamate is called maneb, and the zinc and sodium salts are zineb and nabam, respectively. The iron salt of ethylenebisdithiocarbamate is called ferbam, and the zinc salt of this ion is called ziram. These salts are chelates (Section 5.4) in which two S atoms from the ethylenebis-dithiocarbamate anion are bonded to the same metal ion in a ring structure.

The dithiocarbamate fungicides have been popular for agricultural use because of their effectiveness and relatively low toxicities to animals. However, there is concern over their environmental breakdown products, particularly ethylenethiourea (2-imidazolidine-thione, Figure 12.5), which is toxic to the thyroid and has been shown to be mutagenic, carcinogenic, and teratogenic in experimental animals.

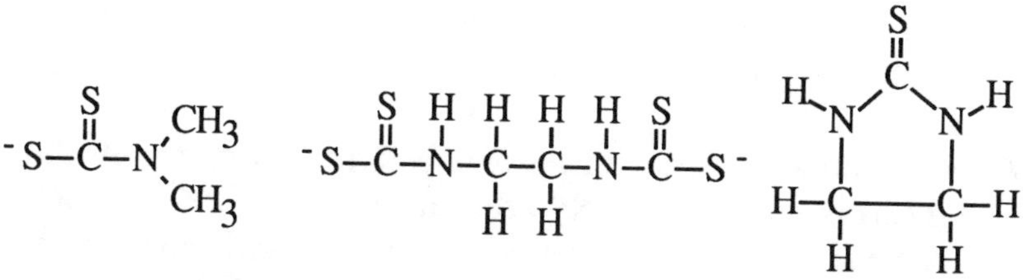

Dimethyldithiocarbamate anion — Ethylenebisdithiocarbamate anion — Ethylenethiourea

Figure 12.5. Dithiocarbamate anions and ethylenethiourea.

12.4. SULFOXIDES AND SULFONES

Numerous important organic compounds contain both sulfur and oxygen. Among these are the **sulfoxides** and **sulfones**, shown by the examples in Figure 12.6.

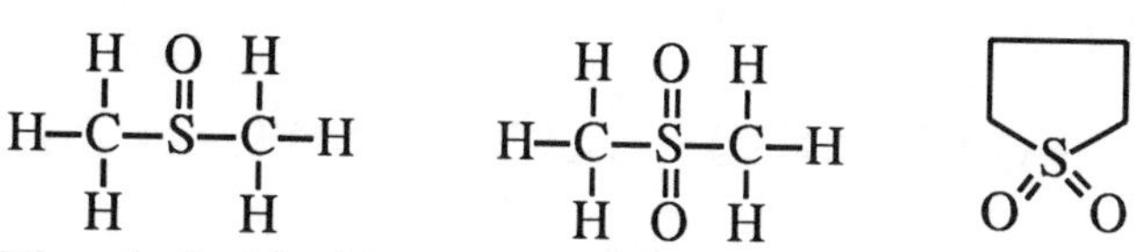

Dimethylsulfoxide (DMSO) — Dimethylsulfone — Sulfolane

Figure 12.6. Sulfoxides and sulfones.

Dimethylsulfoxide (DMSO) is a liquid with numerous uses and some very interesting properties. Mixed with water, it produces a good antifreeze solution. It is also employed to remove paint and varnish and as a hydraulic fluid. It has some potential pharmaceutical applications, for example, as an anti-inflammatory and bacteriostatic agent. It has the ability to carry solutes into the skin's stratum corneum (see Figure 3.1) from which they are slowly released into the blood and lymph system. This phenomenon has some pharmaceutical potential, as well as some obvious hazards. Dimethylsulfoxide has a remarkably low acute toxicity, with an LD_{50} of 10–20 *grams* per kg in several kinds of experimental animals. DMSO applied to the skin rapidly spreads throughout the body, and the subject experiences a taste in the mouth resembling that of garlic and quickly develops a garlic odor in the breath. Some DMSO is excreted directly in the urine and it also undergoes partial metabolism to dimethylsulfide and dimethylsulfone (Figures 12.2 and 12.6, respectively).

Although dimethylsulfone has some commercial uses, **sulfolane** (Figure 12.6) is the most widely used sulfone.[5] It is a polar aprotic (no ionizable H) solvent with a relatively high dielectric constant, and it dissolves both organic and inorganic solutes. When ionic compounds are dissolved in sulfolane, the cations are solvated (bound by the solvent) rather strongly. However, the anions are left in a relatively unsolvated form, which tends to increase their reactivities substantially.[6] The major commercial use of sulfolane is in an operation called BTX processing in which it selectively extracts benzene, toluene, and xylene from aliphatic hydrocarbons. It is also the solvent in the Sulfinol process by which thiols (Section 12.2) and acidic compounds are removed from natural gas. Sulfolane is used as a solvent for polymerization reactions and as a polymer plasticizer. Exposure to sulfolane can cause eye and skin irritation, although its overall toxicity is relatively low.

12.5. SULFONIC ACIDS, SALTS, AND ESTERS

Sulfonic acids contain the $-SO_3H$ group attached to a hydrocarbon moiety. For many applications these acids are converted to salts, such as sodium 1-(*p*-sulfophenyl)decane, a biodegradable detergent surfactant. Its structural formula and those of two sulfonic acids are shown in Figure 12.7.

In general, the sulfonic acids are water-soluble and are strong acids because of virtually complete loss of ionizable H^+ in aqueous solution. They have some important commercial applications, such as in the hydrolysis of fats and oils (see Figure 2.12) to fatty acids and glycerol. Benzenesulfonic acid is fused with NaOH in the preparation of phenol. Dyes and some pharmaceutical compounds are manufactured from *p*-toluenesulfonic acid. Methanesulfonic acid has been developed as an esterification catalyst in place of sulfuric acid for the synthesis of resins in paints and coatings. A strong acid, one of its major advantages over sulfuric acid is that it is not an oxidizing species.[7]

```
   H  H  H  H  O
   |  |  |  |  ||
H--C--C--C--C--S--OH
   |  |  |  |  ||
   H  H  H  H  O
```

Butanesulfonic acid

```
           O
           ||
(C6H5)  -- S--OH
           ||
           O
```

Benzenesulfonic acid

```
   H  H  H  H  H  H  H  H  H  H              O
   |  |  |  |  |  |  |  |  |  |              ||
H--C--C--C--C--C--C--C--C--C--C--(C6H4)--S--O- Na+
   |  |  |  |  |  |  |  |  |  |              ||
   H  H  H  H  H  H  H  H  H  H              O
```

Sodium 1–(*p*–sulfophenyl)decane

Figure 12.7. Sulfonic acids and a sulfonate salt.

Benzenesulfonic acid and *p*-toluenesulfonic acid are strong irritants to skin, eyes, and mucous membranes. Solutions of sulfonic acids are strongly acidic and precautions appropriate to the handling of strong acids should be taken with them.

The methyl ester of methylsulfonic acid is methylmethane sulfonate. Its structural formula is the following:

```
   H  O     H
   |  ||    |
H--C--S--O--C--H   Methylmethane sulfonate
   |  ||    |
   H  O     H
```

Toxicologically, it is notable for being a primary or direct-acting carcinogen that does not require metabolic conversion to act as a carcinogen.[8]

12.6. ORGANIC ESTERS OF SULFURIC ACID

As shown in Figure 12.8, esters of sulfuric acid exist in which either one or both of the ionizable H atoms are replaced by hydrocarbon substituents, such as the methyl group. Replacement of 1 H yields an acid ester and replacement of both yields an ester. Metabolically, acid ester sulfates are synthesized in phase II reactions to produce water-soluble products of xenobiotic compounds (such as phenol) that are readily eliminated from the body (see Section 4.3).

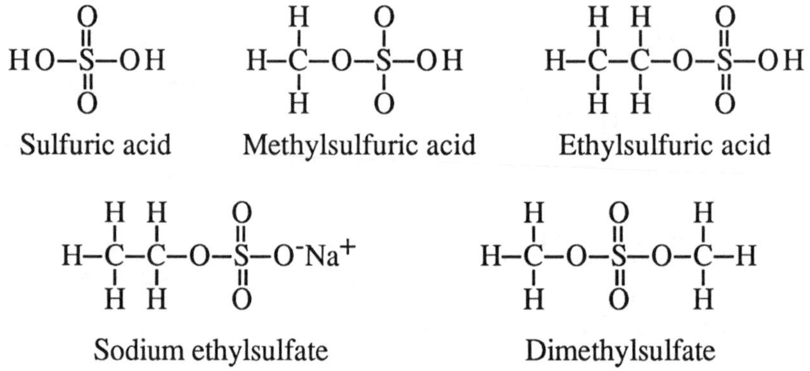

```
     O              H    O                H  H    O
     ‖              |    |                |  |    ‖
HO—S—OH          H—C—O—S—OH           H—C—C—O—S—OH
     ‖              |    |                |  |    ‖
     O              H    O                H  H    O
Sulfuric acid   Methylsulfuric acid    Ethylsulfuric acid

     H  H    O                      H    O    H
     |  |    ‖                      |    ‖    |
  H—C—C—O—S—O⁻Na⁺               H—C—O—S—O—C—H
     |  |    ‖                      |    ‖    |
     H  H    O                      H    O    H
  Sodium ethylsulfate               Dimethylsulfate
```

Figure 12.8. Sulfuric acid and organosulfate esters.

Sulfuric acid esters have several industrial uses, especially as alkylating agents, which act to attach alkyl groups (such as methyl) to organic molecules. Among the products made with sulfuric acid ester reactants are agricultural chemicals, dyes, and drugs.

Methylsulfuric acid is an oily water-soluble liquid. It is a strong irritant to skin, eyes, and mucous tissue. **Ethylsulfuric acid** is likewise an oily liquid and a strong tissue irritant. **Sodium ethylsulfate** is a hygroscopic white crystalline solid.

Dimethylsulfate is a liquid (bp 188°C, fp -32°C). It is colorless, odorless, and highly toxic. It is a primary carcinogen.[8] When skin or mucous membranes are exposed to dimethylsulfate, there is an initial latent period during which few symptoms are observed. After this period, conjunctivitis and inflammation of nasal tissue and respiratory tract mucous membranes develop. Heavier exposures damage the liver and kidney and cause pulmonary edema and cloudiness of the cornea. Death can follow in 3–4 days. The related compound, diethyl-sulfate, is an oily liquid. It reacts with water to yield sulfuric acid. Like dimethylsulfate, it is a strong irritant to tissue and has proven to be carcinogenic in experimental animals.

12.7. MISCELLANEOUS ORGANOSULFUR COMPOUNDS

A number of sulfur compounds containing other elements, such as the halides, are used for various purposes. Some examples of such compounds are shown in Figure 12.9 and discussed briefly here.

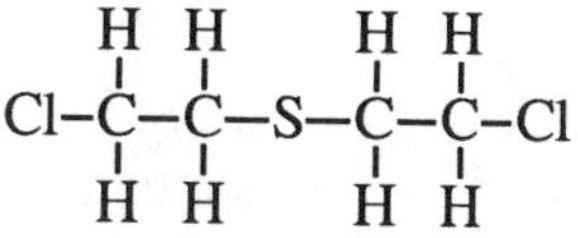

Mustard oil (H)

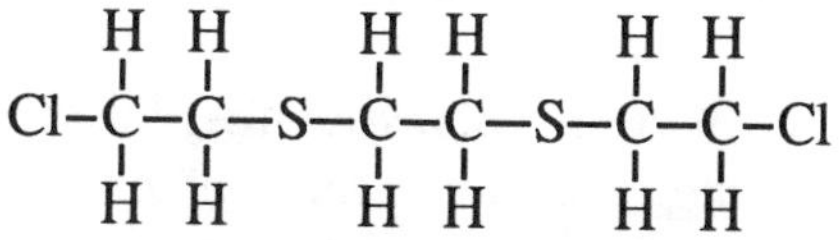

Sesquimustard (Q)

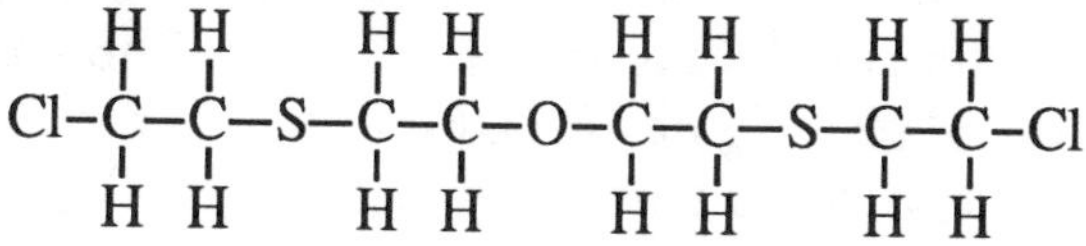

O–Mustard (T)

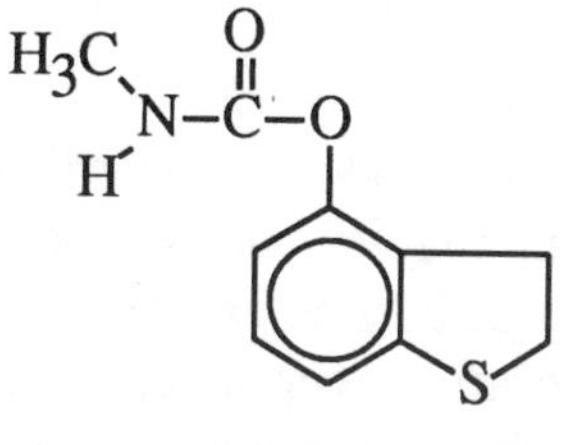

Mobam insecticide

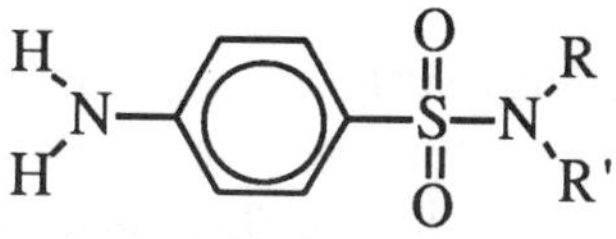

General formula of sulfa drugs

Figure 12.9. Some miscellaneous organosulfur compounds.

Sulfur Mustards

The first three compounds shown in Figure 12.9 are **sulfur mustards**, which are highly toxic military poisons,[9] or "poison gases." These are mustard oil (bis(2-chloroethyl)sulfide), sesquimustard (1,2-bis(2-chloroethylthio)ethane), and O-mustard (bis(2-chloroethylthioethyl)ether). The toxic properties of mustard oil are typical of those of the sulfur mustards. As a military "blistering gas" poison, the vapors of this compound are very penetrating, so that it damages and destroys tissue at some depth from the point of contact. Affected tissue becomes severely inflamed and the resulting lesions often become infected. Death can result from pulmonary lesions. Part of the hazard of mustard oil stems from the speed with which it penetrates tissue, so that efforts to remove it from the exposed area are ineffective after about 30 minutes. The compound is an experimental mutagen[10] and primary carcinogen.[8]

Sulfur in Pesticides

In Section 12.3, rodenticidal thioureas, insecticidal thiocyanates, and fungicidal dithiocarbamates were discussed. Sulfur is a common constituent of other classes of insecticides. These include prominently the organophosphate insecticides discussed in Chapter 13. Mobam (Figure 12.9) is a contact insecticide of the carbamate type, closely related in structure and function to the well-known carbamate insecticide carbaryl (see Figure 10.8). Mobam has been found to have a relatively high toxicity to laboratory mammals and is considerably more toxic than carbaryl.

Sulfa Drugs

The general structure representing sulfa drugs (sulfonamides) is shown in Figure 12.9, where the R groups may be various substituents. In the simplest of these, sulfanilamide, both R groups are H. It was once the most commonly used therapeutic sulfonamide but, because of side effects in humans, is now limited largely to the practice of veterinary medicine. It has a toxicity rating of 3. A large number of therapeutic sulfonamides have been produced. Some of the compounds have a tendency to cause injury to the urinary tract by precipitating in the kidney.

LITERATURE CITED

1. Dreisbach, Robert H., and William O. Robertson, *Handbook of Poisoning*, 12th ed., Appleton and Lange, Norwalk, Conn., 1987.

2. Gosselin, Robert E., Roger P. Smith, and Harold C. Hodge, "ANTU," in *Clinical Toxicology of Commercial Products,* 5th ed., Williams and Wilkins, Baltimore/London, 1984, pp III-40–III-42.

3. Gosselin, Robert E., Roger P. Smith, and Harold C. Hodge, "Sulfathiazole," in *Clinical Toxicology of Commercial Products,* 5th ed., Williams and Wilkins, Baltimore/London, 1984, p. II–394.

4. Murphy, Sheldon D., "Toxic Effects of Pesticides," Chapter 25 in *Casarett and Doull's Toxicology*, 3rd ed., Curtis D. Klaassen, Mary O. Amdur, and John Doull, Eds., Macmillan Publishing Co., New York, 1986, pp. 519–581.

5. Lindstron, Merlin, and Ralph Williams, "Sulfolanes and Sulfones," *Kirk-Othmer Concise Encyclopedia of Chemical Technology*, Wiley-Interscience, New York, 1985, p. 1125.

6. Morrison, Robert T., and Robert N. Boyd, *Organic Chemistry*, Allyn and Bacon, Boston, 1973, pp. 31–32.

7. "Organotopics," Pennwalt, Inc., *Chemical and Engineering News*, July 27, 1987, pp. 43–46.

8. Levi, Patricia E., "Toxic Action," Chapter 6 in *Modern Toxicology*, Ernest Hodgson and Patricia E. Levi, Eds., Elsevier, New York, 1987, pp. 133–184.

9. "Global Experts Offer Advice for Chemical Weapons Treaty," *Chemical and Engineering News*, July 27, 1987, pp. 16–17.

10. Sax, N. Irving, *Dangerous Properties of Industrial Materials*, 6th ed., Van Nostrand Reinhold, New York, 1984.

13

Phosphorus-Containing Organic Compounds

13.1. INTRODUCTION

Phosphorus is directly below nitrogen in the periodic table. (The relationship of the chemistry of phosphorus to that of nitrogen is somewhat like the sulfur-oxygen relationship discussed in the introduction to Chapter 12.) The phosphorus atom electron configuration is {Ne}$3s^2 3p^3$, and it has five outer-shell electrons as shown by its Lewis symbol in Figure 13.1. Because of the availability of underlying 3*d* orbitals, the valence shell of phosphorus can be expanded to more than 8 electrons.

$$\cdot\dot{\underset{\cdot}{\mathrm{P}}}: \qquad\qquad \begin{array}{c}\mathrm{H}\\ :\ddot{\underset{\cdot\cdot}{\mathrm{P}}}:\mathrm{H}\\ \mathrm{H}\end{array}$$

Lewis symbol of phosphorus atom — Lewis structural formula of phosphine

Figure 13.1. Lewis representations of the phosphorus atom and its hydride, phosphine, showing valence electrons as dots.

There are many kinds of organophosphorus compounds, including those with P-C bonds and those in which hydrocarbon moieties are bonded to P through an atom other than carbon, usually oxygen. These compounds have numerous industrial uses and many of them, especially the organophosphate ester insecticides (see Section 13.7), are economic poisons. Organophosphorus compounds have varying degrees of toxicity. Some of these compounds, such as the "nerve gases" produced as industrial poisons, are deadly in minute quantities.

Phosphine

Phosphine (PH_3) is the hydride of phosphorus discussed as a toxic inorganic compound in Section 7.10. The formulas of many organophosphorus compounds can be derived by substituting organic groups for the H atoms in phosphine, and such an approach serves as a good starting point for the discussion of organophosphorus compounds.

13.2. ALKYL AND ARYL PHOSPHINES

Figure 13.2 gives the structural formulas of the more significant alkyl and aryl phosphine compounds. **Methylphosphine** is a colorless reactive gas that is very toxic by inhalation. **Dimethylphosphine** is a colorless, reactive, volatile liquid (bp 25°C). It is toxic by inhalation and ingestion. Both methylphosphine and dimethylphosphine have toxic effects similar to those of phosphine, a pulmonary tract irritant and central nervous system depressant that causes fatigue, vomiting, difficult breathing and even death. **Trimethylphosphine** is a colorless volatile liquid (bp 42°C). It is reactive enough to be spontaneously ignitable and probably has a high toxicity. **Triethylphosphine** probably has a high toxicity and tributylphosphine is a moderately toxic liquid. **Phenylphosphine** (phosphaniline) is a reactive, moderately flammable liquid (bp 16°C) with a high toxicity by inhalation. **Triphenylphosphine** is a crystalline solid (mp 79°C, bp > 360°C) with a low reactivity and moderate toxicity when inhaled or ingested.

The combustion of aryl and alkyl phosphines, such as trimethylphosphine, occurs as shown by the following example:

$$4C_3H_9P + 26O_2 \rightarrow 12CO_2 + 18H_2O + P_4O_{10} \quad (13.1)$$

Such a reaction produces P_4O_{10}, a corrosive irritant toxic substance discussed in Section 7.10, or droplets of corrosive orthophosphoric acid, H_3PO_4.

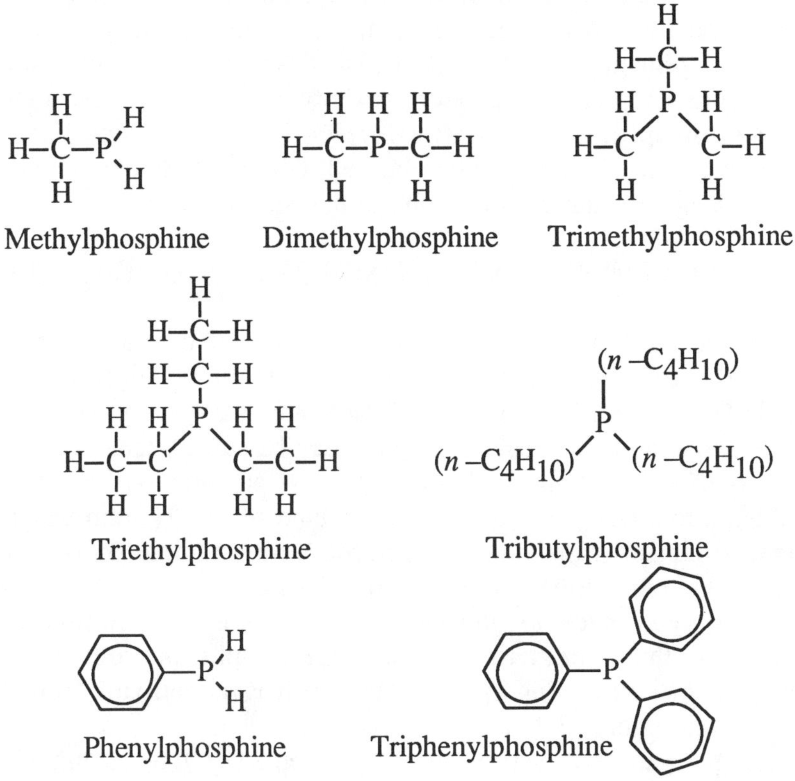

Methylphosphine Dimethylphosphine Trimethylphosphine

Triethylphosphine

$(n\text{-}C_4H_{10})$ P $(n\text{-}C_4H_{10})$ $(n\text{-}C_4H_{10})$

Tributylphosphine

Phenylphosphine Triphenylphosphine

Figure 13.2. Some of the more significant alkyl and aryl phosphines.

13.3. PHOSPHINE OXIDES AND SULFIDES

Phosphine oxides and sulfides have the general formulas illustrated below, where the Rs represent hydrocarbon groups:

$$R{-}\overset{\overset{\large O}{\|}}{\underset{\underset{\large R'}{|}}{P}}{-}R''$$

Phosphine oxide

$$R{-}\overset{\overset{\large S}{\|}}{\underset{\underset{\large R'}{|}}{P}}{-}R''$$

Phosphine sulfide

Two common phosphine oxides are **triethylphosphine oxide** (each R is a C_2H_5 group) and **tributylphosphine oxide** (each R is a C_4H_9 group). The former is a colorless, deliquescent, crystalline solid (mp 52.9°C, bp 243°C). The latter compound is a crystalline solid (mp 94°C). Both compounds probably have high toxicities when ingested.

Triethylphosphine sulfide, $(C_2H_5)_3PS$, is a crystalline solid (mp 94°C). Not much is known about its toxicity, which is probably high. **Tributylphosphine sulfide**, $(C_4H_9)_3PS$, is a skin irritant with a moderate toxicity hazard.[1] When burned, both of these compounds give off dangerous fumes of phosphorus and sulfur oxides.

13.4 PHOSPHONIC AND PHOSPHOROUS ACID ESTERS

Phosphonic acid esters are derived from phosphonic acid (often erroneously called phosphorous acid), which is shown with some of its esters in Figure 13.3. Only two of the H atoms of phosphonic acid are ionizable, and hydrocarbon groups may be substituted for these atoms to give phosphonic acid esters. It is also possible to have esters in which a hydrocarbon moiety is substituted for the H atom that is bonded directly to the phosphorus atom. An example of such a compound is **benzylphosphonic acid**, diethyl ester, shown in Figure 13.3. This type of compound has the same elemental formula as triesters of the hypothetical acid, $P(OH)_3$, phosphorous acid. Examples of triesters of phosphorous acid such as **trimethylphosphite** are shown in Figure 13.3.

Trimethylphosphite is a colorless liquid (bp 233°C). It is soluble in many organic solvents, but not in water. Little information is available regarding its toxicity or other hazards. **Tributylphosphite** is a liquid, bp 120°C. It decomposes in water, but is probably not very toxic. **Triphenylphosphite** is a white solid or oily liquid (mp 23°C, bp 157°C). It is a skin irritant with a moderate oral toxicity. Although it is not soluble in water, it may hydrolyze somewhat to phenol, which adds to its toxicity. **Tris(2-ethylhexyl)phosphite** (a trialkyl phosphite in which the hydrocarbon moieties are the 2-ethylhexyl group, $-CH_2CH(C_2H_5)C_4H_9$) is a water-insoluble compound (bp 100°C). Its toxicity is largely unknown.

Methylphosphonate $(CH_3O)P(O)H(OH)$ has a moderate oral toxicity and is a skin and eye irritant. **Dibutylphosphonate** (formula $(C_4H_9O)_2P(O)H$) is a liquid boiling at 115°C at 10 mm Hg pressure. Through ingestion and dermally it has a moderately high toxicity. Like other organophosphonates and phosphites, it can decompose to evolve dangerous products when heated, burned, or exposed to reactive chemicals, such as oxidants. Thermal decomposition can result in the evolution of highly toxic phosphine, PH_3. Combustion produces corrosive orthophosphoric acid and oxides of phosphorus.

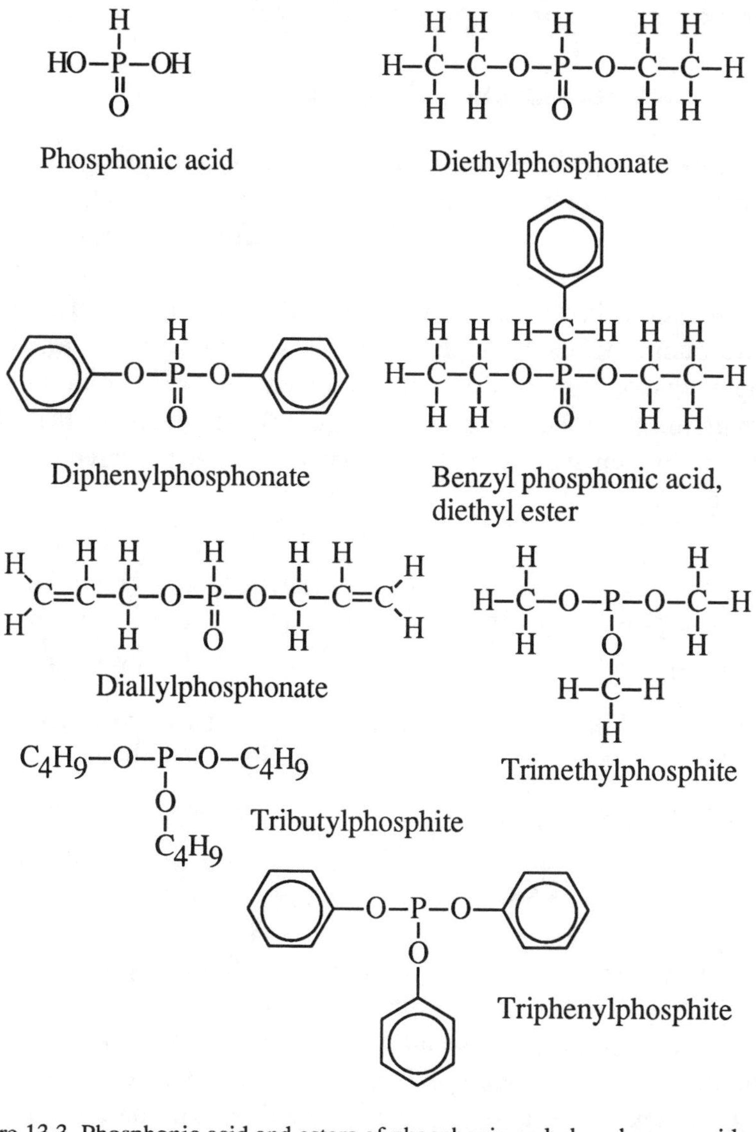

Phosphonic acid

Diethylphosphonate

Diphenylphosphonate

Benzyl phosphonic acid, diethyl ester

Diallylphosphonate

Trimethylphosphite

Tributylphosphite

Triphenylphosphite

Figure 13.3 Phosphonic acid and esters of phosphonic and phosphorous acids.

Diallylphosphonate, shown in Figure 13.3, has two alkenyl substituent groups. Information is lacking on its toxicity, although compounds with allyl groups tend to be relatively toxic. Incidents have been reported in which this compound has exploded during distillation.[2]

13.5. ORGANOPHOSPHATE ESTERS

Orthophosphates and Polyphosphates

Figure 13.4 shows the structural formula of orthophosphoric acid as well as those of diphosphoric and polyphosphoric acids produced by polymerization of orthophosphoric acid with loss of water. These compounds form esters in which alkyl, alkenyl, and aryl hydrocarbon moieties are substituted for H; most of the more common ones are esters of orthophosphoric acid. In this section only the relatively simple organophosphate esters are discussed. Many economic poisons — particularly insecticides — are organophosphate esters that often contain nitrogen and sulfur. These compounds are discussed in a later section.

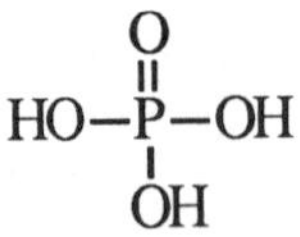

$HO-P(=O)(OH)-OH$ Orthophosphoric acid

$HO-P(=O)(OH)-O-P(=O)(OH)-OH$ Pyrophosphoric acid

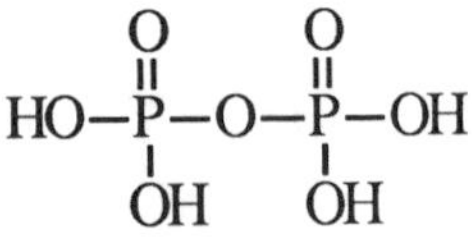

$HO-P(=O)(OH)-[O-P(=O)(OH)]_n-O-P(=O)(OH)-OH$ Polyphosphoric acids (n = 3 and higher)

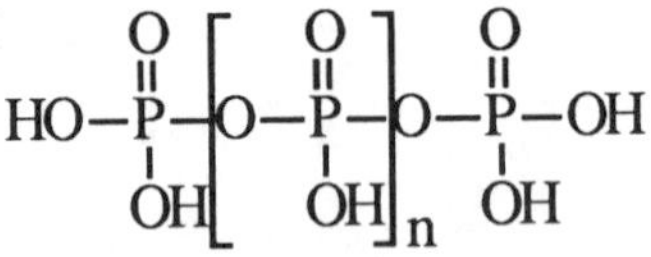

Figure 13.4. Orthophosphoric acid and acids formed by its polymerization.

Orthophosphate Esters

Some of the more significant phosphate esters are shown in Figure 13.5. Trimethylphosphate is the simplest of the organophosphate esters; the structural formulas of the other alkyl esters of orthophosphoric acid are like those of trimethylphosphate, but with alkyl substituent groups other than methyl. Comparatively little information is available about the toxicity of trimethylphosphate, although it is probably moderately toxic orally or through skin absorption.

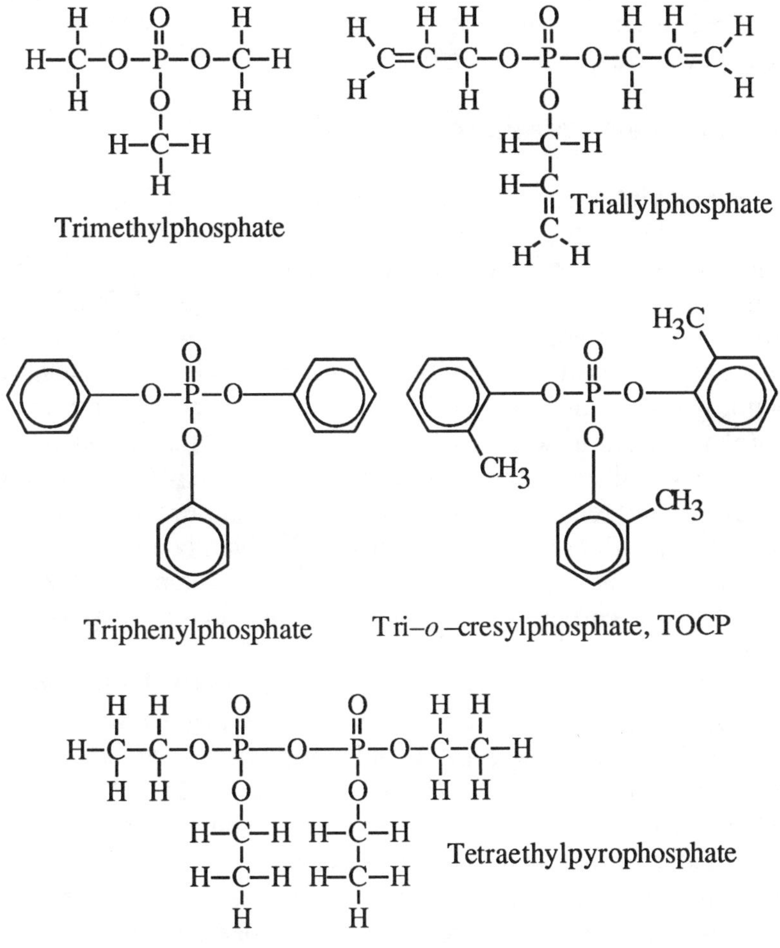

Trimethylphosphate

Triallylphosphate

Triphenylphosphate

Tri-*o*-cresylphosphate, TOCP

Tetraethylpyrophosphate

Figure 13.5. Phosphate esters.

Triethylphosphate, $(C_2H_5O)_3PO$, is a liquid (fp -57°C, bp 214°C). It is insoluble in water, but soluble in most organic solvents. Like other phosphate esters, it damages nerves and is a cholinesterase inhibitor. It is regarded as moderately toxic. Two other alkyl phosphates with toxicities probably similar to that of triethyphosphate are **tributylphosphate**, $(n\text{-}C_4H_9O)_3PO$, and **tris(2-ethylhexyl)phosphate**, $(C_8H_{17}O)_3PO$.

Triallylphosphate is the phosphate triester of allyl alcohol and contains unsaturated C=C bonds in its structure. This compound is a liquid (fp -50°C). It is regarded as having a high toxicity and produces abnormal tissue growth when administered subcutaneously. It has been known to explode during distillation.

Aromatic Phosphate Esters

Triphenylphosphate is a colorless, odorless, crystalline solid (mp 49°C, bp 245°C). It is moderately toxic. A similar, but much more toxic, compound is **tri-*o*-cresyl-phosphate, TOCP,** an aryl phosphate ester with a notorious record of poisonings.[3] Before its toxicity was fully recognized, TOCP was a common contaminant of commercial **tricresylphosphate**. Tricresylphosphate is an industrial chemical with numerous applications and consists of a mixture of phosphate esters in which the hydrocarbon moieties are *meta* and *para* cresyl substituents. It has been used as a lubricant, gasoline additive, flame retardant, solvent for nitrocellulose, plasticizer, and even a cooling fluid for machine guns. Although modern commercial tricresylphosphate contains less than 1% TOCP, contaminant levels of up to 20% in earlier products have resulted in severe poisoning incidents.

Pure TOCP is a colorless liquid (fp -27°C, bp 410°C). It produces pronounced neurological effects and causes degeneration of the neurons in the body's central and peripheral nervous systems, although fatalities are rare. Early symptoms of TOCP poisoning include nausea, vomiting, and diarrhea accompanied by severe abdominal pain. Normally a 1–3 week latent period occurs after these symptoms have subsided, followed by manifestations of peripheral paralysis as evidenced by "wrist drop" and "foot drop." In some cases, the slow recovery is complete, whereas in others partial paralysis remains.

The most widespread case of TOCP poisoning occurred in the U.S. in 1930 when approximately 20,000 people were affected by the ingestion of alcoholic Jamaican ginger ("Jake") adulterated by 2% TOCP. The peculiar manner in which the victims walked, including "foot drop," slapping the feet on the floor, high stepping, and unsteadiness, gave rise to the name of "jake-leg" to describe the very unfortunate condition.

A major incident of TOCP poisoning affected 10,000 people in Morocco in 1959. The victims had eaten food cooked in olive oil adulterated with TOCP-contaminated lubricating oil. A number of cases of permanent paralysis resulted from ingestion of the contaminated cooking oil.

It is believed that metabolic products of TOCP inhibit acetylcholinesterase. Apparently other factors are involved in its neurotoxicity. Despite the devastating effects of TOCP, the percentage of virtually complete recovery in healthy subjects is relatively high.

Tetraethylpyrophosphate

Tetraethylpyrophosphate, TEPP, was the first organophosphate compound to be used as an insecticide.[4] This compound was developed in Germany during World War II and was substituted for nicotine as an insecticide. It is a white to amber hygroscopic liquid (bp 155°C) that readily hydrolyzes in contact with water. Because of its tendency to hydrolyze and its extremely high toxicity to mammals, TEPP was used for only a very short time as an insecticide, although it is a very effective one. It was typically applied as an insecticidal dust formulation containing 1% TEPP.

The toxicity of TEPP to humans and other mammals is very high and it has a toxicity rating of 6, supertoxic. TEPP is a very potent acetylcholinesterase inhibitor. (The inhibition of acetylcholinesterase by organophosphate insecticides is discussed in Section 13.7.)

13.6. PHOSPHOROTHIONATE AND PHOSPHORODITHIOATE ESTERS

The general formulas of **phosphorothionate** and **phosphorodithioate** esters are shown in Figure 13.6, where R represents a hydrocarbon or substituted hydrocarbon moiety. Many of the organophosphate insecticides are sulfur-containing esters of these general types, which often exhibit higher insect:mammal toxicity ratios than do their non-sulfur analogs. Esters containing the P=S (thiono) group are not as effective as their analogous compounds that contain the P=O functional group in inhibiting acetylcholinesterase.[5] In addition to their lower toxicities to non-target organisms, thiono compounds are more stable toward non-enzymatic hydrolysis. The metabolic conversion of P=S to P=O (oxidative desulfuration) in organisms is responsible for the insecticidal activity and mammalian toxicity of phosphorothionate and phosphorodithioate insecticides.

```
        S                    S
        ‖                    ‖
R—O—P—O—R          R—O—P—S—R
        |                    |
        O                    O
        |                    |
        R                    R
```

Phosphorothionate   Phosphorodithioate

Figure 13.6. General formulas of phosphorothionate and phosphorodithioate esters; each R represents a hydrocarbon or substituted hydrocarbon moiety.

An example of a simple phosphorothionate is tributylphosphorothionate, in which the R groups (above) are n-C_4H_9 groups. It is a colorless liquid (bp 143°C). The compound is a cholinesterase inhibitor, as are some of its metabolic products. Examples of phosphorothionate and phosphorodithioate esters with more complex formulas synthesized for their insecticidal properties are discussed in the following section.

13.7. ORGANOPHOSPHATE INSECTICIDES

The organophosphate insecticides were originally developed in Germany during the 1930s and 1940s, primarily through the efforts of Gerhard Schrader and his research group. The first of these was tetraethylpyrophosphate, TEPP, discussed in Section 13.5. Its disadvantages — including high toxicity to mammals — led to the development of related compounds, starting with **parathion**, *O,O*-diethyl-*O*-*p*-nitrophenylphosphorothionate, which will be discussed in some detail.

Chemical Formulas and Properties

Many insecticidal organophosphate compounds have been synthesized. Unlike the organohalide insecticides that they largely displaced, the organophosphates readily undergo biodegradation and do not bioaccumulate. The most common organophosphate insecticides can be represented by the general formulas for phosphorothionate esters (Figure 13.7), phosphorodithioate esters (Figure 13.8), and their oxygen analogs, phosphate esters (Figure 13.9). In these generalized formulas, R is a methyl ($-CH_3$) or ethyl ($-C_2H_5$) group and Ar is a moiety of more complex structure, frequently aromatic. Example organophosphate insecticides based upon these three types of esters are also shown in these figures.

Phosphorothionate Insecticides

Figure 13.7 gives the structural formulas of three typical phosphorothionate esters and the general formula of this type of organophosphate insecticide.

General formula

Parathion

Diazinon

Chlorothion

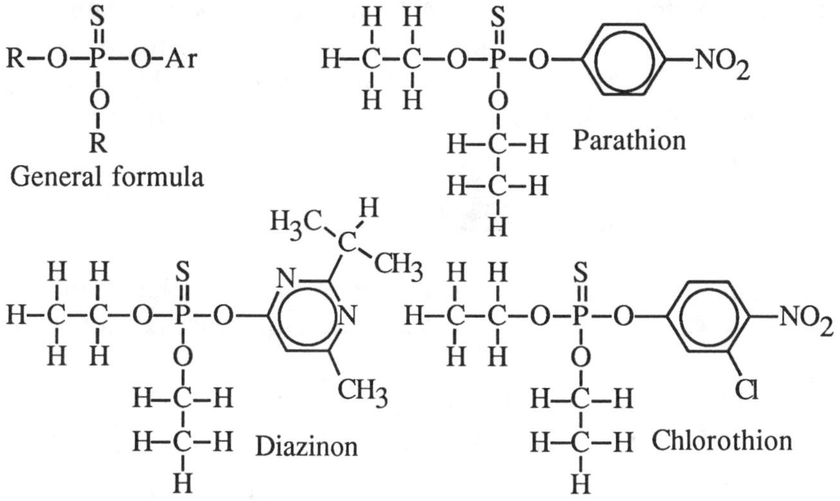

Figure 13.7. Phosphorothionate organophosphate insecticides.

Parathion

Insecticidal parathion is a phosphorothionate ester first licensed for use in 1944. Pure parathion is a yellow liquid that is insoluble in kerosene and water, but stable in contact with water. Among its properties that make parathion convenient to use as an insecticide are stability in contact with neutral and somewhat basic aqueous solutions, low volatility, and toxicity to a wide range of insects. It is applied as an emulsion in water, dust, wettable powder, or aerosol. It is not recommended for applications in homes or animal shelters because of its toxicity to mammals.

Parathion has a toxicity rating of 6 (supertoxic) and methylparathion (which has methyl groups instead of the ethyl groups shown in Figure 13.7) is regarded as extremely toxic.[6] As little as 120 mg of parathion has been known to kill an adult human and a dose of 2 mg has killed a child. Most accidental poisonings have occurred by absorption through the skin. Since its use began, several hundred people have been killed by parathion. One of the larger poisoning incidents occurred in Jamaica in 1976 from ingestion of parathion-contaminated flour. Of 79 people exposed, 17 died.

In the body, parathion is converted to paraoxon (structure in Figure 13.9), which is a potent inhibitor of acetylcholinesterase. Because this conversion is required for parathion to have a toxic effect, symptoms develop several hours after exposure, whereas the toxic effects of TEPP or paraoxon develop much more rapidly. Symptoms of parathion poisoning in humans include skin twitching, respiratory distress, and, in fatal cases, respiratory failure due to central nervous system paralysis.

Phosphorodithioate Insecticides

Figure 13.8 shows the general formula of phosphorodithioate insecticides and structural formulas of some examples, of which

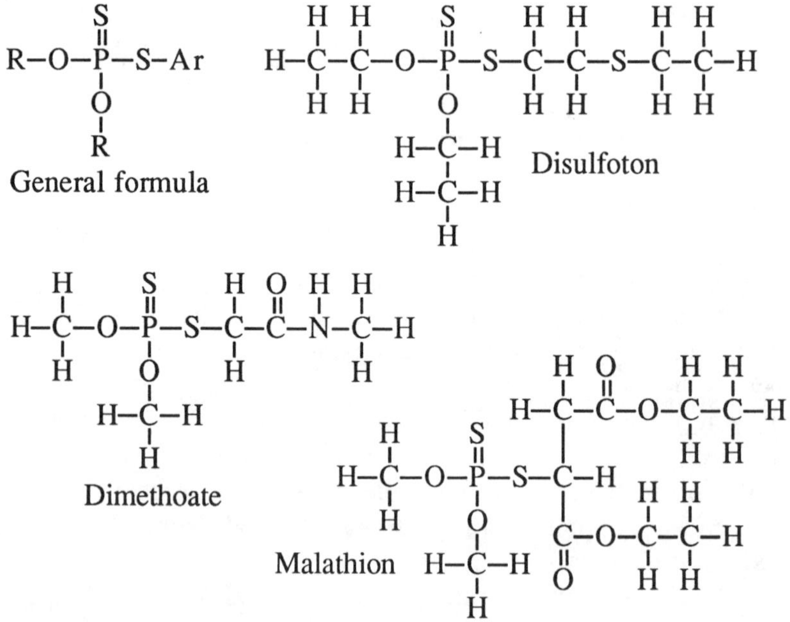

```
      S                  H H     S     H H     H H
      ||                 | |     ||    | |     | |
R-O-P-S-Ar           H-C-C-O-P-S-C-C-S-C-C-H
      |                  | |     |     | |     | |
      O                  H H     O     H H     H H
      |                          |
      R                        H-C-H     Disulfoton
General formula                  |
                               H-C-H
                                 |
                                 H

   H     S     H O   H
   |     ||    | ||  |
H-C-O-P-S-C-C-N-C-H
   |     |     |     |
   H     O     H     H
         |
       H-C-H
         |
         H
   Dimethoate

                                     H O     H H
                                     | ||    | |
                                   H-C-C-O-C-C-H
                    H     S          |       | |
                    |     ||         |       H H
                  H-C-O-P-S-C-H
                    |     |          |     H H
                    H     O          |     | |
                          |          C-O-C-C-H
       Malathion        H-C-H        ||    | |
                          |          O     H H
                          H
```

Figure 13.8. Phosphorodithioate organophosphate insecticides.

malathion is the best known. Malathion shows how differences in structural formula can cause pronounced differences in the properties of organophosphate pesticides. Malathion has two carboxyester linkages which are hydrolyzable by carboxylase enzymes to relatively non-toxic products as shown by the following reaction:

```
                 H  O      H  H
                 |  ||     |  |
               H-C--C--O-- C--C-H
   H     S       |         |  |
   |     ||      |         H  H     H2O, Carboxylesterase
 H-C--O--P--S--C-H      H  H       --------------------->
   |     |       |       |  |       enzyme
   H     O       C--O--C--C-H
         |       ||    |  |
       H-C-H     O     H  H
         |
         H
                         H  O
                         |  ||
                       H-C--C-OH
         H      S        |                 H  H
         |      ||       |                 |  |
       H-C--O---P--S-- C-H    +  2HO--C--C-H   (13.2)
         |      |        |                 |  |
         H      O        C-OH              H  H
                |        ||
              H-C-H      O
                |
                H
```

The enzymes that accomplish this reaction are possessed by mammals, but not by insects, so that mammals can detoxify malathion, whereas insects cannot. The result is that malathion has selective insecticidal activity. For example, although malathion is a very effective insecticide, its LD50 for adult male rats is about 100 times that of parathion, reflecting the much lower mammalian toxicity of malathion compared to some of the more toxic organophosphate insecticides, such as parathion.

Carboxylase enzymes are inhibited by organophosphates other than malathion. The result of exposure of mammals to malathion plus another organophosphate is potentiation (see Section 3.7) of the toxicity of malathion.

Phosphate Ester Insecticides

Figure 13.9 shows some organophosphate insecticides based upon the phosphate esters. These compounds do not contain sulfur. One of the more significant of these compounds is paraoxon which, as noted previously, is a metabolic activation product of parathion. It has been synthesized directly and was made by Schrader in 1944 along with parathion. One of the most toxic organophosphate insecticides, paraoxon has a toxicity rating of 6. **Mevinphos** is considered to be an extremely dangerous chemical. **Dichlorvos** has a toxicity rating of 4 and is deactivated by enzymes in the livers of mammals. Its tendency to vaporize has enabled its use in "pest strips."

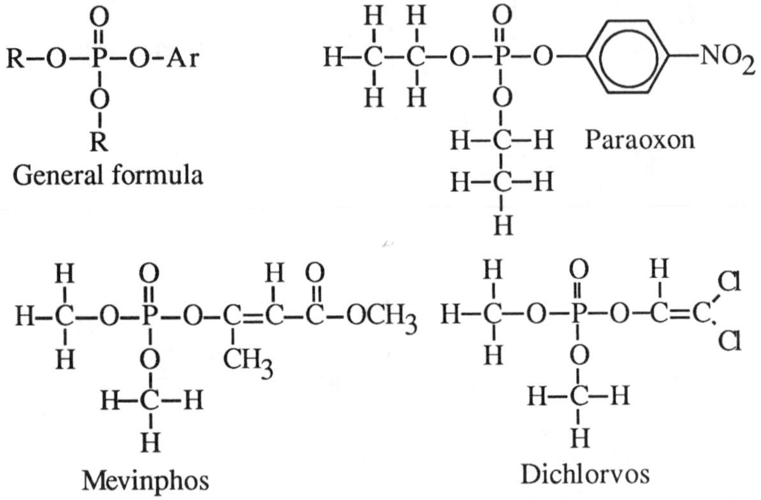

```
        O
        ||
R—O—P—O—Ar
        |
        O
        |
        R
General formula
```

```
   H  H      O
   |  |      ||
H—C—C—O—P—O—⟨benzene ring⟩—NO2
   |  |      |
   H  H      O
             |
          H—C—H     Paraoxon
             |
          H—C—H
             |
             H
```

```
   H      O        H  O
   |      ||       |  ||
H—C—O—P—O—C=C—C—OCH3
   |      |      |
   H      O      CH3
          |
       H—C—H
          |
          H
     Mevinphos
```

```
   H      O      H   Cl
   |      ||     |  /
H—C—O—P—O—C=C
   |      |        \
   H      O         Cl
          |
       H—C—H
          |
          H
     Dichlorvos
```

Figure 13.9. Organophosphate insecticides based on phosphate esters.

Toxic Actions of Organophosphate Insecticides

Inhibition of Acetylcholinesterase

The organophosphate insecticides inhibit acetylcholinesterase in mammals and insects.[7] As discussed in Section 1.10, acetylcholine forms during the transmission of nerve impulses in the body, including the central nervous system, and it must be hydrolyzed by the action of acetylcholinesterase enzyme to prevent excessive stimulation of the nerve receptors. Accumulation of acetylcholine can cause numerous effects related to excessive nerve response. Among these effects in humans are bronchioconstriction resulting in chest tightness and wheezing; stimulation of muscles in the intestinal tract resulting in nausea, vomiting and diarrhea; and muscular twitching and cramps. The central nervous system shows numerous effects from the accumulation of acetylcholine. These include psychological symptoms of restlessness, anxiety, and emotional instability. The subject may suffer from headache and insomnia. In more severe cases, depression of the respiratory and circulatory systems, convulsions, and coma may result. In fatal poisonings, death is due to respiratory system paralysis.

Cholinesterase inhibition occurs when an inhibitor, I, binds to the cholinesterase enzyme, E, to produce an enzyme-inhibitor complex as shown by the following reaction:[7]

$$E + I \rightleftarrows EI \qquad (13.3)$$

With some inhibitors the reaction is reversible. With other kinds of compounds, such as the organophosphates, a stable, covalently bound complex, E', is formed from which it is difficult to regenerate the original enzyme as illustrated by the reaction

$$E + I \rightleftarrows EI \rightarrow E' \xrightarrow[\text{at all}]{\text{Slowly, or not}} E + \text{Products} \qquad (13.4)$$

An example of irreversible binding is that of paraoxon, which can be viewed as an organophosphate compound containing a phosphorylating group, P, and a leaving group, L, as shown below:

Bond cleaved

```
    H  H     O
    |  |     ||
 H–C–C–O–P–O–⟨benzene ring⟩–NO2
    |  |     |
    H  H     O
             |
           H–C–H
             |
           H–C–H
             |
             H
```

Leaving group

Phosphorylating group

The reaction of this compound with cholinesterase enzyme, E, can be represented by the following reaction:

$$E + PL \rightarrow EP + L$$

$$\xrightarrow[\text{bound enzyme}]{\text{Slow dissociation of covalently}} E + \text{Products} \qquad (13.5)$$

The phosphorylating group bonds to an OH group at the active site of the enzyme.

Metabolic Activation

Highly purified phosphorothionate and phosphorodithioate insecticides do not inhibit acetylcholinesterase directly. In order for these compounds to inhibit acetylcholinesterase, the following phase I metabolic conversion of P=S to P=O must occur:

$$\text{R–O–}\overset{\overset{\displaystyle S}{\|}}{\underset{\underset{\displaystyle R}{|}}{\underset{\displaystyle O}{\underset{|}{P}}}}\text{–O–Ar} + \{O\} \xrightarrow{\text{Metabolic oxidation}} \text{R–O–}\overset{\overset{\displaystyle O}{\|}}{\underset{\underset{\displaystyle R}{|}}{\underset{\displaystyle O}{\underset{|}{P}}}}\text{–O–Ar} \quad (13.6)$$

A specific example of this type of reaction is the conversion of parathion to paraoxon mentioned in the preceding section.

Mammalian Toxicities

The mammalian toxicities of the organophosphate insecticides vary widely. This may be seen from the LD_{50} values to male rats of organophosphate insecticides, including some of the ones discussed in the preceding section. Where the approximate LD_{50} values (oral, mg/kg) are given in parentheses, a listing of common organophosphate insecticides in descending order of toxicity is TEPP (1) > mevinphos, disulfoton (6-7) > parathion, methylparathion, azinphos-methyl, chlorfenvinphos (13-15) > dichlorvos (80) > diazinon (110) > trichlorfon (215) > chlorothion (880) > ronnel > malathion (1300).

Deactivation of Organophosphates

The deactivation of organophosphates is accomplished by hydrolysis as shown by the following general reactions where R is an alkyl group, Ar is a substituent group that is frequently aromatic, and X is either S or O:

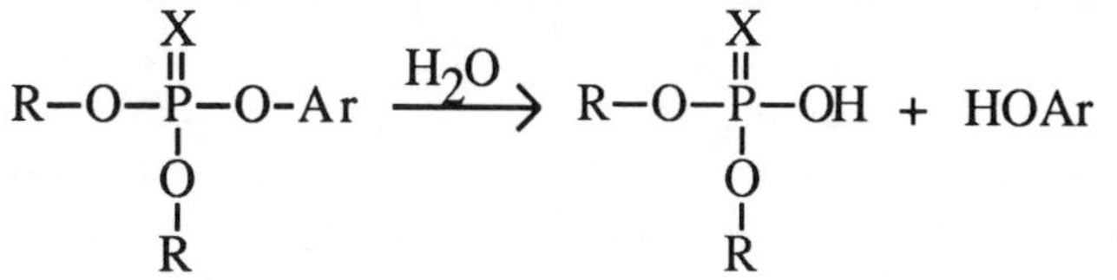

$$\mathrm{R{-}O{-}\overset{\displaystyle X}{\overset{\|}{P}}{-}O{-}Ar}\ (\text{with } \mathrm{O{-}R} \text{ on P}) \xrightarrow{H_2O} \mathrm{R{-}O{-}\overset{\displaystyle X}{\overset{\|}{P}}{-}OH}\ (\text{with } \mathrm{O{-}R} \text{ on P}) + \mathrm{HOAr} \quad (13.7)$$

$$\mathrm{R{-}O{-}\overset{\displaystyle X}{\overset{\|}{P}}{-}O{-}Ar}\ (\text{with } \mathrm{O{-}R} \text{ on P}) \xrightarrow{H_2O} \mathrm{R{-}O{-}\overset{\displaystyle X}{\overset{\|}{P}}{-}OAr}\ (\text{with } \mathrm{OH} \text{ on P}) + \mathrm{HOR} \quad (13.8)$$

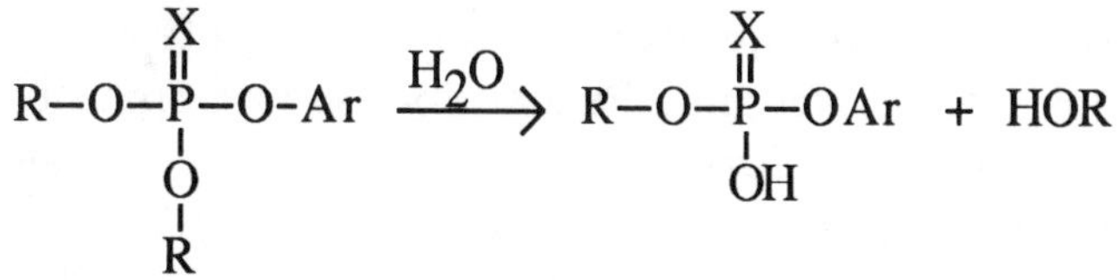

13.8. ORGANOPHOSPHORUS MILITARY POISONS

Organophosphorus compounds developed for use as military poisons — the "nerve gases" — are among the most toxic synthetic compounds ever made.[8] Two example structures of these compounds are shown in Figure 13.10. Other chemical warfare agents containing organic phosphorus include **Tabun** (O-ethyl N,N-dimethylphosphoramidocyanidate), **Soman** (*o*-pinacolyl methylphosphonofluoridate), and **"DF"** (methylphosphonyldifluoride).

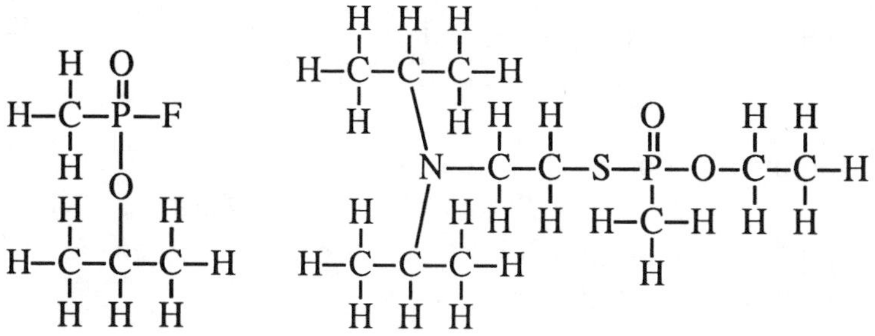

```
                     H  H  H
                     |  |  |
   H  O           H--C--C--C--H
   |  ||             |  |
H--C--P--F           H  \ H  H  H       O      H  H
   |  |                  \   |  |       ||     |  |
   H  |                   N--C--C--S--P--O--C--C--H
      O                  /   |  |       |      |  |
   H  |  H           H  / H  H  H    H--C--H   H  H
   |  |  |           |  |  |            |
H--C--C--C--H     H--C--C--C--H         H
   |  |  |           |  |  |
   H  H  H           H  H  H
    Sarin                    VX
```

Figure 13.10. Two examples of organophosphate military poisons.

The action of **Sarin** is typical of the organophosphorus military poisons. Its lethal dose to humans may be as low as about 0.01 mg/kg. It is a systemic poison to the central nervous system that is readily absorbed as a liquid through the skin; a single drop so absorbed can kill a human. It is a colorless liquid (fp -58°C, bp 147°C). Pure Tabun is also a colorless liquid; it has a freezing point of -49°C and decomposes when heated to 238°C. Its toxicity is similar to that of Sarin. Tabun acts primarily on the sympathetic nervous system and it has a paralytic effect on the blood vessels. Its toxic action and symptoms of poisoning are similar to those of parathion, an organophosphate insecticide for which extensive human toxicity data are available. **Diisopropyl fluorophosphate**, $(i\text{-}C_3H_7O)_2P(O)F$, is a highly toxic oily liquid that served as the basis for the development of "nerve gases" in Germany during World War II. The organophosphorus military poisons are powerful inhibitors of acetylcholinesterase enzyme.

LITERATURE CITED

1. Sax, N. Irving, *Dangerous Properties of Industrial Materials,* 6th ed., Van Nostrand Co., New York, 1984.

2. *Chemistry Laboratory Safety Library,* 491 M, 5th ed., National Fire Protection Association, Boston, Mass., 1975.

3. Gosselin, Robert E., Roger P. Smith, and Harold C. Hodge, "Tri-*ortho*-cresyl Phosphate," in *Clinical Toxicology of Commercial Products,* 5th ed., Williams and Wilkins, Baltimore/London, 1984, pp. III-388–III-393.

4. Murphy, Sheldon D., "Toxic Effect of Pesticides," Chapter 18 in *Casarett and Doull's Toxicology,* 3rd ed., Curtis D. Klaassen, Mary O. Amdur, and John Doull, Eds., Macmillan Publishing Co., New York, 1986, pp. 519–581.

5. Hodgson, Ernest, "Metabolism of Toxicants," Chapter 3 in *A Textbook of Modern Toxicology,* Ernest Hodgson and Patricia E. Levi, Eds., Elsevier, New York, 1987, pp. 51-84.

6. Gosselin, Robert E., Roger P. Smith, and Harold C. Hodge, "Parathion," in *Clinical Toxicology of Commercial Products,* 5th ed., Williams and Wilkins, Baltimore/London, 1984, pp. III-336–III-333.

7. Mann, A. Russell, "Cholinesterase Inhibitors," Chapter 11 in *Introduction to Biochemical Toxicology*, Ernest Hodgson and Frank E. Guthrie, Eds., Elsevier, New York, 1986, pp. 194–223.

8. "Global Experts Offer Advice for Chemical Weapons Treaty," *Chemical and Engineering News*, July 27, 1987, pp. 16–17.

14

Toxic Natural Products

14.1. INTRODUCTION

Toxic natural products are poisons produced by organisms. They include an enormous variety of materials. Perhaps the most acutely toxic substance known is botulism toxin produced by the anaerobic bacterium *Clostridium botulinum*, and responsible for many food poisoning deaths, especially from improperly canned food. Mycotoxins generated by fungi (molds) can cause a number of human maladies, and some of these materials, such as the aflatoxins, are carcinogenic to some animals. Venoms from wasps, spiders, scorpions, and reptiles can be fatal to humans. Each year, in the Orient, tetrodotoxin from improperly prepared puffer fish makes this dish the last delicacy consumed by some unfortunate diners. The stories of Socrates' execution from being forced to drink an extract of the deadly poisonous spotted hemlock plant and Cleopatra's suicide at the fangs of a venomous asp are rooted in antiquity.

Living organisms wage chemical warfare against their potential predators and prey with a fascinating variety of chemical substances. Some organisms produce toxic metabolic by-products that have no obvious use to the organisms that make them. This chapter briefly describes some of the toxic natural products from living organisms with emphasis on those that are toxic to humans.

One distinction should be made at this point that applies particularly to animals. A **poisonous organism** is one that produces toxins. A **poisonous animal** may contain toxins in its tissues that act as poisons to other animals that eat its flesh. **Venoms** are poisons that can be delivered without the need for the organism to be eaten. **Venomous animals** can deliver poisons to another animal by means such as biting (usually striking with fangs) or stinging. The puffer fish — some tissues of which are deadly when ingested — is a poisonous animal, whereas the rattlesnake is a venomous one, although its flesh may be eaten safely. Some organisms — notoriously, the skunk — wage chemical warfare by emitting substances that are not notably toxic, but still effective in keeping predators away. Perhaps these organisms should be classified as noxious species.

14.2. TOXIC SUBSTANCES FROM BACTERIA

The two greatest concerns regarding toxic substances from bacteria are their roles in causing symptoms of bacterial disease and food poisoning. It is useful to consider as one class those bacteria that produce toxins that adversely affect a host in which the bacteria are growing and, as another class, bacteria that produce toxins to which another organism is subsequently exposed, such as by ingestion.

Bacteria are single-celled microorganisms that may grow in colonies and are shaped as spheres, rods, or spirals. They are usefully classified with respect to their need for oxygen, which accepts electrons during the metabolic oxidation of food substances, such as organic matter. **Aerobic bacteria** require molecular oxygen to survive, whereas **anaerobic bacteria** grow in the absence of oxygen, which may be toxic to them. **Facultative bacteria** can grow either aerobically or anaerobically. Anaerobic bacteria and facultative bacteria functioning anaerobically use substances other than molecular O_2 as electron acceptors (oxidants that "take electrons away" from other reactants in a chemical reaction). For example, sulfate takes the place of O_2 in the anaerobic degradation of organic matter (represented as $\{CH_2O\}$) by *Desulfovibrio*, yielding toxic hydrogen sulfide (H_2S) as a product as shown by the following overall reaction.[1]

$$SO_4^{2-} + 2\{CH_2O\} + 2H^+ \rightarrow H_2S + 2CO_2 + 2H_2O \qquad (14.1)$$

Toxicants such as H_2S produced by microorganisms are usually not called microbial toxins, a term that more properly refers to usually proteinaceous species of high molecular mass synthesized metabolically by microorganisms and capable of inducing a strong response in susceptible organisms at low concentrations. Bacteria and other microorganisms do produce a variety of poisonous substances, such as acetaldehyde, formaldehyde, and putrescine (see Figure 10.2).

In Vivo Bacterial Toxins

Some important bacterial toxins are produced in the host and have a detrimental effect on the host. For example, such toxins are synthesized by *Clostridium tetani*, common soil bacteria that enter the body largely through puncture wounds.

The toxin from this bacterium interferes with neurotransmitters, such as acetylcholine, causing **tetanus**, commonly called lockjaw. Abnormal populations or strains of *Shigella dysenteriae* bacteria in the body can cause a severe form of dysentery because they release a toxin that causes intestinal hemorrhaging and gastrointestinal tract paralysis. Toxin-releasing bacteria responsible for the most common form of food poisoning are those of the genus *Salmonella*. Victims are afflicted with flu-like symptoms and may even die from the effects of the toxin. Diphtheria is caused by a toxin generated by *Cornybacterium diphtheriae*. The toxin interferes with protein synthesis and is generally destructive to tissue.

Bacterial Toxins Produced Outside the Body

The most notorious toxin produced by bacteria outside the body is that of *Clostridium botulinum*. This kind of bacteria grows naturally in soil and on vegetable material. Under anaerobic or slightly aerobic conditions it synthesizes an almost unbelievably toxic product. The conditions for generating this toxin most commonly occur as the result of the improper canning of food, particularly vegetables. Botulinum toxin binds irreversibly to nerve terminals, preventing the release of acetylcholine; the affected muscle acts as though the nerve were disconnected. The toxin actually consists of several polypeptides in the range of 200,000 to 400,000 molecular mass. Fortunately, these proteins are inactivated by heating for a sufficient time at 80–100°C. Botulinum poisoning symptoms appear within 12–36 hours after ingestion, beginning with gastrointestinal tract disorders and progressing through neurologic symptoms, paralysis of the respiratory muscles, and death by respiratory failure.

14.3. MYCOTOXINS

Mycotoxins are toxic metabolites from fungi that have a wide range of structures and a variety of toxic effects.[2] Perhaps the most well known of these are the **aflatoxins** produced by *Aspergillus*. Another major class of mycotoxins consists of the **ergot alkaloids** from *Claviceps*. Several genera of *fungi imperfecti* produce toxic tricothecenes.

Aflatoxins

The most common source of aflatoxins is moldy food, particularly nuts, some cereal grains, and oil seeds. The most notorious of the aflatoxins is aflatoxin B_1, for which the structural formula is shown in Figure 14.1. Produced by *Aspergillus niger*, it is a potent liver toxin and liver carcinogen in some species. It is metabolized in the liver to an epoxide (see Section 4.2). The product is electrophilic with a strong tendency to bond covalently to protein, DNA, and RNA. Other common aflatoxins produced by molds are those designated by the letters B_2, G_1, G_2, and M_1.

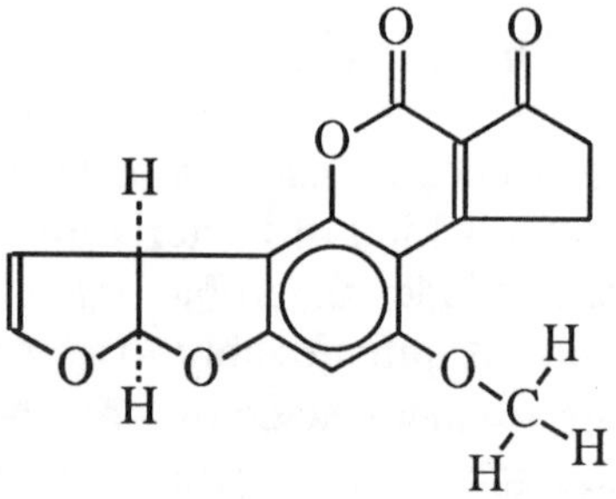

Aflatoxin B_1

Figure 14.1. Structural formula of aflatoxin B_1, a mycotoxin.

Other Mycotoxins

The ergot alkaloids have been associated with a number of spectacular outbreaks of central nervous system disorders, sometimes called ergotism. St. Anthony's fire is an example of convulsive ergotism. From examination of historical records, it is now known that outbreaks of this malady resulted for the most part from ingestion of moldy grain products. Although ergotism is now virtually unknown in humans, it still occurs in livestock.

Trichothecenes are composed of 40 or more structurally related compounds produced by a variety of molds, including *Cephalosporium*, *Fusarium*, *Myrothecium*, and *Trichoderma*,[3] which grow predominantly on grains. Much of the available information on human toxicity of trichothecenes was obtained from an outbreak of poisoning in Siberia in 1944. During the food shortages associated with World War II, the victims ate moldy barley, millet, and wheat. People who ate this grain suffered from skin inflammation; gastrointestinal tract disorders, including vomiting and diarrhea; and multiple hemorrhage. About 10 percent of those afflicted died.

14.4. TOXINS FROM PROTOZOA

Toxic substances from two of the major types of unicellular protista — bacteria and fungi — were discussed in the preceding sections. Another type of protista notable for the production of toxic substances consists of the **protozoans**. These organisms are generally regarded as single-celled animals. Most of the protozoans that produce toxins belong to the order **dinoflagellata**, which are predominantly marine species. The cells of these organisms are enclosed in cellulose envelopes, which often have beautiful patterns on them.[4] Among the effects caused by toxins from these organisms are gastrointestinal, respiratory, and skin disorders in humans; mass kills of various marine animals; and paralytic conditions caused by eating infested shellfish.

The marine growth of dinoflagellates is characterized by occasional incidents in which they multiply at such an explosive rate that they color the water yellow, olive-green, or red by their vast numbers. In 1946, some sections of the Florida coast became so afflicted by "red tide" that the water became viscous and for many miles the beaches were littered with the remains of dead fish, shellfish, turtles, and other marine organisms. The sea spray in these areas became so irritating that coastal schools and resorts were closed.

The greatest danger to humans from dinoflagellata toxins comes from the ingestion of shellfish, such as mussels and clams, that have accumulated the protozoa from sea water. In this form the toxic material is called paralytic shellfish poison. As little as 4 mg of this toxin, the amount found in several severely infested mussels or clams, can be fatal to a human. The toxin depresses respiration and affects the heart, resulting in complete cardiac arrest in extreme cases.

14.5. TOXIC SUBSTANCES FROM PLANTS

Various plants produce a wide range of toxic substances as reflected by plant names such as "deadly nightshade" and "poison hemlock." Although the use of "poison arrows" having tips covered with plant-derived curare has declined as the tribes that employed them have acquired the sometimes dubious traits of modern civilization, poisoning by plants is still of concern in the grazing of ruminant animals, and houseplants such as philadendron and yew are responsible for poisoning some children. Plant-derived cocaine causes many deaths among those who use it or get into fatal disputes marketing it.

Toxic substances from plants are discussed here in the five categories of nerve poisons, internal organ poisons, skin and eye irritants, allergens, and metal (mineral) accumulators. As can be seen in Figure 14.2, plant toxins have a variety of chemical structures. Prominent among the chemical classes of toxicants synthesized by plants are nitrogen-containing alkaloids (see Section 10.9) that usually occur in plants as salts. Some harmful plant compounds undergo metabolic reactions to form toxic substances. For example, amygdalin, present in the meats of fruit seeds such as those of apples and peaches, undergoes acid hydrolysis in the stomach or enzymatic hydrolysis elsewhere in the body to yield toxic HCN.

Nerve Toxins

Nerve toxins from plants cause a variety of central nervous and peripheral nervous system effects. Several examples are cited here.

Plant-derived **neurotoxic psychodysleptics** affect peripheral neural functions and motor coordination, sometimes accompanied by delirium, stupor, trance states, and vomiting.[5] Prominent among these toxins are the pyrollizidines from peyote. Also included are erythrionones from the coral tree and quinolizidines from the mescal bean (see Figure 14.2).

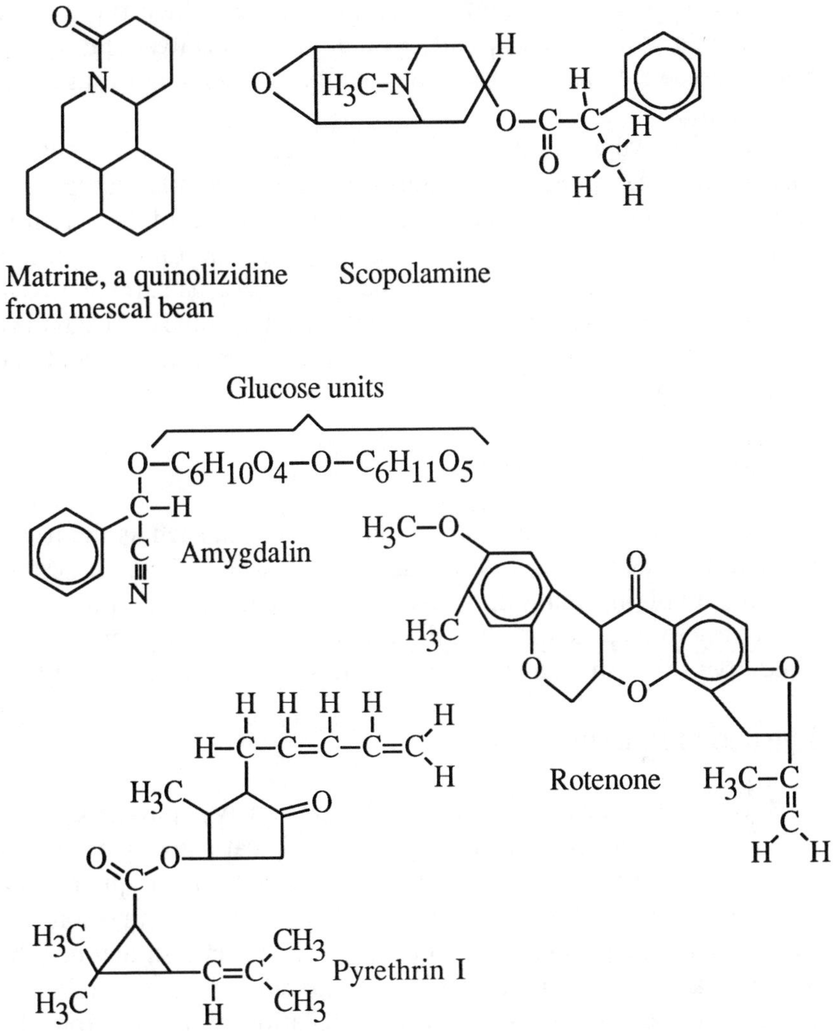

Matrine, a quinolizidine from mescal bean

Scopolamine

Glucose units

Amygdalin

Rotenone

Pyrethrin I

Figure 14.2. Some representative toxic substances from plants.

Spotted hemlock contains the alkaloid nerve toxin coniine (see Figure 10.10). Ingestion of this poison is followed within about 15 minutes by symptoms of nervousness, trembling, arrythmia, and bradycardia. Body temperature may decrease and fatal paralysis can occur. The Nightshade family of plants contains edible potato, tomato, and eggplant. However, it also contains the "deadly nightshade," or *Atropa Belladonna* (beautiful woman). This toxic plant contains scopolamine (Figure 14.2) and atropine. Ingestion causes dizziness, mydriasis, speech loss, and delirium. Paralysis can occur. Fatally poisoned victims may expire within half an hour of ingesting the poison.

Several nerve toxins produced by plants are interesting because of their insecticidal properties. Among them are six similar compounds called **pyrethrins**, and including pyrethrin I (Figure 14.2), pyrethrin II, jasmolin I and II, and cinerin I and II. The chrysanthemum is a pyrethrin source. Insecticidal nicotine (Figure 3.2) is extracted from tobacco. Rotenone (Figure 14.2) is synthesized by almost 70 legumes. This insecticidal compound is safe for most mammals, with the notable exception of swine.

Internal Organ Plant Toxins

Toxins from plants may affect internal organs, such as the heart, kidney, liver, and stomach. Because of their much different digestive systems involving multiple stomachs, ruminant animals may react differently to these toxins than do monogastric animals. Some major plant toxins that affect internal organs are summarized in Table 14.1.

Eye and Skin Irritants

Anyone who has been afflicted by poison ivy, poison oak, or poison sumac appreciates the high potential that some plant toxins have to irritate skin and eyes. The toxic agents in the plants just mentioned are catechol compounds, such as urishikiol in poison ivy. Contact with the poison causes a characteristic skin rash that may be disabling and very persistent in heavily exposed, sensitive individuals. Lungs may be affected — often by inhalation of smoke from the burning plants — to the extent that hospitalization is required.

Photosensitizers constitute a class of systemic plant poisons capable of affecting areas other than those exposed. These pigmented substances may pass through the liver without being conjugated and collect in skin capillaries. When these areas of the skin are subsequently exposed to light, the capillaries leak. (Since light is required, the phenomenon is called a photosensitized condition.) In severe cases, tissue and hair are sloughed off. St. John's wort or horsebrush causes this kind of condition in farm animals.

Table 14.1. Plant Toxins that Affect Internal Organs of Animals

Toxin	Source	Target organ
Saponins	Alfalfa Cockles English ivy	Noncardioactive steroid glycosides that cause gastric upset
Pyrrolizidine alkaloids such as festucine	Fescue hay	Liver: obstructs veins
Hypercin	St. John's wort, horse-brush	Liver: releases pigmented molecules to the bloodstream causing photosensitization
Digitoxin	Foxglove	Heart: overdose causes heart to stop, but lower doses are used to strengthen heartbeat and eliminate fluids present in congestive heart failure.
Oxalates	Oak tannin	Kidney: precipitation of CaC_2O_4 obstructs kidney tubules.

Allergens

Many plants are notorious for producing allergens that cause allergic reactions in sensitized individuals. The most common plant allergens consist of pollen. The process that leads to an allergic reaction starts when the allergen, acting as a hapten, combines with an endogenous protein in the body to form an antigen. Antibodies are generated that react with the antigen and produce histamine, resulting in an allergic reaction. (See Section 3.10, Immune System Response.) The severity of the symptoms varies with the amount of histamine produced. These symptoms can include skin rash, watery eyes, and running nose. In severe cases, victims suffer fatal anaphylactic shock. An example of an allergic reaction to a plant product — all too familiar to many of its victims — is hay fever induced by the pollen of ragweed or goldenrod.

Mineral Accumulators

Some plants classified as mineral accumulators become toxic because of the inorganic materials that they absorb from soil and water and retain in the plant biomass. An important example of such a plant is *Astragalus*, sometimes called "locoweed." This plant causes serious problems in some western U.S. grazing areas because it accumulates selenium. Animals that eat too much of it get selenium poisoning, characterized by anemia and a condition known descriptively as "blind staggers."

Nitrate accumulation may occur in plants growing on soil fertilized with nitrate under moisture-deficient conditions, such as those that occurred in the the central U.S. during the record-setting spring/summer drought of 1988. In the stomachs of ruminant animals, nitrate (NO_3^-) ingested with plant material is reduced to nitrite (NO_2^-). The nitrite product enters the bloodstream and oxidizes the iron(II) in hemoglobin to iron(III). The condition that results is methemoglobinemia, which was discussed in Section 10.3 in connection with aniline poisoning.

Another toxicological problem that can result from excessive nitrate in plant material is the generation of toxic nitrogen dioxide gas (see Section 7.3) during the fermentation of ensilage composed of chopped plant matter contaminated with nitrate. The toxic effect of NO_2 from this source has been called "silo-filler's disease."

Mushroom Toxins

Although mushrooms are fungal bodies, their toxic effects are often discussed along with those of toxic plants. Some mushrooms, such as *Amanita phalloides*, *Amanita virosa*, and *Gyromita esculenta*, are very toxic, with reported worldwide deaths of the order of 100 per year.[6] In extreme cases, one bite of one poisonous mushroom can be fatal. Accidental mushroom poisonings are often caused by the Death's Head mushroom, because it is easily mistaken for edible varieties.

Some toxins in mushrooms are alkaloids that cause central nervous system effects of narcosis and convulsions. Hallucinations occur in subjects who have eaten mushrooms that contain psilocybin. The toxic alkaloid muscarine is present in some mushrooms.

Another class of toxins produced by some mushrooms consists of polypeptides, particularly amanitin and phalloidin. These substances are stable to heating (cooking). They are systemic poisons that attack cells of various organs, including the heart and liver. In early 1988 an organ transplant was performed on a woman in the U.S. to replace her liver, which was badly damaged from the ingestion of wild mushrooms that she and a companion had mistakenly collected and consumed as edible varieties.

The symptoms of mushroom poisoning vary. Typical early symptoms involve the gastrointestinal tract and include stomach pains and cramps, nausea, vomiting, and diarrhea. Victims in the second phase of severe poisoning may suffer paralysis, delirium, and coma along with often severe liver damage.

Edible *Coprinus atramentarius* mushrooms produce an interesting ethanol-sensitizing effect similar to that of disulfiram (antabuse, see Section 12.3).[7] Ingestion of alcohol can cause severe reactions in individuals up to several days after having eaten this kind of mushroom.

14.6. INSECT TOXINS

Although relatively few insect species produce enough toxin to endanger humans, insects cause more fatal poisonings in the U.S. each year than do all other venomous animals combined. Most venomous insects are from the order *Hymenoptera*, which includes ants, bees, hornets, wasps, and yellowjackets. These insects deliver their toxins by a stinging mechanism.

Chemically, the toxic substances produced by insects are variable and have been incompletely characterized. In general, *Hymenopteran* venoms are composed of water-soluble, nitrogen-containing chemical species in concentrated mixtures. Although they contain chemical compounds in common, the compositions of insect venoms from different species are variable. The three major types of chemical species are biologically synthesized (biogenic) amines, peptides and small proteins, and enzymes. Of the biogenic amines, the most common is histamine, which is found in the venoms of bees, wasps, and hornets. Wasp and hornet venoms contain serotonin, and hornet venom contains the biogenic amine acetylcholine. Among the peptides and low-molecular-mass proteins in insect venoms are apamin, mellitin, and mast cell degranulating peptide in bee venom; wasp kinin, and hornet kinin. Enzymes contained in bee, wasp, and hornet venom are phospholipase A and hyaluronidase. Phospholipase B occurs in wasp and hornet venom.

Bee Venom

Bee venom contains a greater variety of proteinaceous materials than do wasp and hornet venoms. Apamin in bee venom is a polypeptide containing 18 amino acids and having three disulfide (–SS–) bridges in its structure. Because of these bridges and its small size, the apamine molecule is able to traverse the blood-brain barrier and function as a central nervous system poison. Mellitin in bee venom consists of a chain of 27 amino acids. It can be a direct cause of erythrocyte hemolysis. Symptoms of bradycardia and arrythmia can be caused by mellitin. Mast cell degranulating peptide in bee venom acts on mast cells. These are a type of "white blood cell" believed to be involved in the production of heparin, a key participant in the blood clotting process. The degranulating peptide causes mast cells to disperse, with an accompanying release of histamine into the system.

Wasp and Hornet Venoms

Wasp and hornet venoms are distinguished from bee venoms by their lower content of peptides. They do contain kinin peptide, which may cause smooth muscle contraction and lowered blood pressure. Two biogenic amines in wasp or hornet venoms (serotonin and acetylcholine) lower blood pressure and cause pain. Acetylcholine may cause malfunction of heart and skeletal muscles.

Toxicities of Insect Venoms

The toxicities of insect venoms are low to most people. Despite this, relatively large numbers of fatalities occur each year from insect stings because of allergic reactions in sensitized individuals. These reactions can lead to potentially fatal anaphylactic shock, which affects the nervous system, cardiovascular function, and respiratory function. The agents in bee venom that are responsible for severe allergic reactions are mellitin and two enzymes of high molecular mass — hyaluronidase and phospholipase A-2.

14.7. SPIDER TOXINS

There are about 30,000 species of spiders, virtually all of which produce venom![8] Fortunately, most lack dangerous quantities of venom, or the means to deliver it. Nevertheless, about 200 species of spiders are significantly poisonous to humans. Many of these have colorful common names, such as tarantula, trap-door spider, black widow, giant crab spider, poison lady, and deadly spider. Space permits only a brief discussion of spider venoms here.

Brown Recluse Spiders

Brown recluse spiders (*Loxosceles*) are of concern because of their common occurrence in households in temperate regions. Many people are bitten by this spider despite its non-aggressive nature. A brown spider bite can cause severe damage at the site of the injury. When this occurs, the tissue and underlying muscle around the bite undergo severe necrosis, leaving a gaping wound up to 10 cm across. Plastic surgery is often required in an attempt to repair the damage. In addition, *Loxosceles* venom may cause systemic effects, such as fever, vomiting, and nausea. In rare cases death results. The venom of *Loxosceles* contains protein and includes enzymes. The mechanisms by which the venom produces lesions are not completely understood.

Widow Spiders

The widow spiders are *Latrodectus* species. Unlike the *Loxosceles* species described above, the bite sites from widow spiders show virtually no damage. The symptoms of widow spider poisoning are many and varied. They include pain, cramps, sweating, headache, dizziness, tremor, nausea, vomiting, and elevated blood pressure. The venom contains several proteins, including a proteinaceous neurotoxin with a molecular mass of about 130,000.[9]

Other Spiders

Several other types of venomous spiders should be mentioned here. Running spiders (*Chiranthium* species) are noted for the tenacity with which they cling to the bite area, causing a sharply painful wound. The bites of cobweb spiders (*Steatoda* species) cause localized pain and tissue damage. Venomous jumping spiders (*Phidippus* species) produce a wheal (raised area) up to 5 cm across in the bite area.

14.8. REPTILE TOXINS

Snakes are the most notorious of the venomous animals. The names of venomous snakes suggest danger — Eastern diamondback rattlesnake, king cobra, black mamba, fer-de-lance, horned puff adder, *Crotalus horridus horridus*. About 10 percent of the approximately 3500 snake species are sufficiently venomous to be hazardous to humans.[10] These may be divided among *Crotilidae* (including rattlesnakes, bushmaster, and fer-de-lance), *Elipidae* (including cobras, mambas, and coral snakes), *Hydrophidae* (true sea snakes), *Laticaudae* (sea kraits), and *Colubridae* (including the boomslang and Australian death adder).

Chemical Composition of Snake Venoms

Snake venoms are complex mixtures that may contain biogenic amines, carbohydrates, and metal ions. The most important snake venom constituents, however, are proteins, including numerous enzymes. Among the most prominent of the enzymes in snake venom are the proteolytic enzymes, which bring about the breakdown of proteins, thereby causing tissue to deteriorate. Some proteolytic enzymes are associated with hemorrhaging. Collagen (connective tissue in tendons, skin, and bones) is broken down by collogenase enzyme contained in some snake venoms. Among the other kinds of enzymes that occur in snake venoms are phospholipase enzymes, phosphoesterase enzymes, and acetylcholinesterase.

Numerous non-enzyme polypeptides occur in snake venom. Some of these polypeptides, though by no means all, are neurotoxins.

Toxic Effects of Snake Venom

The effects of snake bite can range from relatively minor discomfort to almost instant death. The latter is often associated with drastically lowered blood pressure and shock. The predominant effects of snake venoms can be divided into the two major categories of cardiotoxic and neurotoxic effects. Blood clotting mechanisms may be affected by enzymes in snake venom, and blood vessels may be damaged as well.[2] Almost all organs in the victims of poisoning by *Crotilidae* have exhibited adverse effects, many of which appear to be associated with changes in the blood and with alterations in the lung. Clumped blood cells and clots in blood vessels have been observed in the lungs of victims. Agents in cobra toxin break down the blood-brain barrier by disrupting capillaries and cell membranes. So altered, the barrier loses its effectiveness in preventing the entry of other brain-damaging toxic agents.

14.9. NON-REPTILE ANIMAL TOXINS

Several major types of animals that produce poisonous substances have been considered so far in this chapter. With the exception of birds, all classes of the animal kingdom contain members that produce toxic substances.[8] Those not covered so far in this chapter are summarized here.

Numerous kinds of fishes contain poisons in their organs and flesh. The most notorious of the poisonous fish are "puffers" or "puffer-like" fish that produce tetrodotoxin. This supertoxic substance is present in the liver and ovary of the fish. It acts on nerve cell membranes by affecting the passage of sodium ions, a process involved in generating nerve impulses. The fatality rate for persons developing clinical symptoms of tetrodotoxin poisoning is about 40 percent. Usually associated with Japan, puffer fish poisoning kills about 100 people per year globally. Some of these poisonings are self-inflicted.

Some fishes are venomous and have means of delivering venom to other animals. This is accomplished, for example, by spines on weever fish. The infamous stingray has a serrated spine on its tail that can be used to inflict severe wounds, while depositing venom from specialized cells along the spine. The venom increases the pain from the wound and has systemic effects, especially on the cardiovascular system.

Numerous species of amphibia (frogs, toads, newts, salamanders) produce poisons, such as bufotenin, in specialized skin secretory glands. Most of these animals pose no hazard to humans. However, some of the toxins are extremely poisonous. For example, Central American Indian hunters have used hunting arrows tipped with poison from the golden arrow frog.

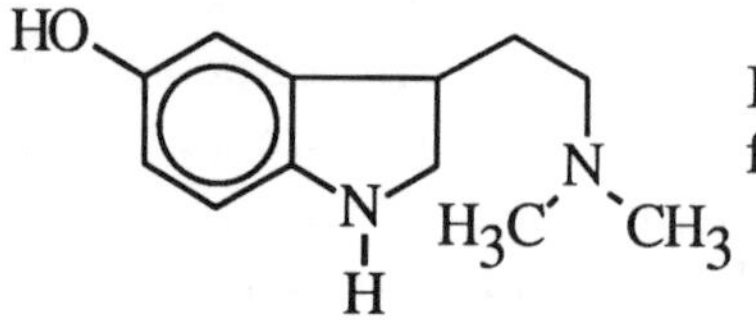

Bufotenin, a compound isolated from some amphibian toxins

In addition to the poisonous fishes and sea snakes mentioned previously in this chapter, several other forms of marine life produce toxins. Among these are *Porifera*, or sponges, consisting of colonies of unicellular animals. The sponges release poisons to keep predators away. They may have sharp spicules that can injure human skin, while simultaneously exposing it to poison. Various species of the *Coelenterates*, including corals, jellyfish, and sea anenomes, are capable of delivering venom by stinging. Some of these venoms have highly neurotoxic effects. *Echinoderms*, exemplified by starfish, sea cucumbers, and sea urchins, may possess spines capable of delivering toxins. People injured by these spines often experience severe pain and other symptoms of poisoning. Various molluscs produce poisons, such as the poison contained in the liver of the abalone, *Haliotis*. Some mollusc poisons, such as those of the genus *Conus*, are delivered as venoms by a stinging mechanism.

Some arthropods other than spiders (see Section 14.7) produce toxins. Prominent among these are the scorpions, whose stings are a very serious hazard, especially to children. In Mexico, the dangerous scorpion *Centruroides suffusus* attains a length up to 9 cm. Mexico has had a serious problem with fatal scorpion bites.

Some centipedes are capable of delivering venom by biting. The site becomes swollen, inflamed, and painful. Some millipedes secrete a toxic skin irritant when touched.

Although the greater hazard from ticks is their ability to carry human diseases, such as Rocky Mountain spotted fever or Lyme disease (a debilitating condition that had become of great concern in some areas during the summer of 1988), some species discharge a venom that causes a condition called tick paralysis, characterized by weakness and lack of coordination. An infamous mite larva, the chigger, causes inflamed spots on the skin that itch badly. The chigger is so small that most people require a magnifying glass to see it, but a large number of chigger bites can cause intense misery in a victim.

LITERATURE CITED

1. Manahan, Stanley E., *Environmental Chemistry*, 4th ed., Brooks/Cole Publishing Co., Pacific Grove, CA, 1984.

2. Hodgson, Ernest, and Patricia E. Levi, *Modern Toxicology*, Elsevier, New York, 1987.

3. Hayes, Johnnie R., and T. Colin Campbell, "Food Additives and Contaminants," Chapter 24 in *Casarett and Doull's Toxicology*, 3rd ed., Curtis D. Klaassen, Mary O. Amdur, and John Doull, Eds., Macmillan Publishing Co., New York, 1986, pp. 771–800.

4. "Dinoflagellida," in *Van Nostrand's Scientific Encyclopedia*, 5th ed., Van Nostrand Reinhold Co., New York, 1976, pp. 804–805.

5. Emboden, William, *Narcotic Plants*, Macmillan Publishing Co., New York, 1979.

6. Dreisbach, Robert H., and William O. Robertson, *Handbook of Poisoning*, 12th ed., Appleton and Lange, Norwalk, Conn., 1987.

7. Lampe, Kenneth F., "Toxic Effects of Plant Toxins," Chapter 23 in *Casarett and Doull's Toxicology*, 3rd ed., Curtis D. Klaassen, Mary O. Amdur, and John Doull, Eds., Macmillan Publishing Co., New York, 1986, pp. 757–767.

8. Russell, Findlay E., "Toxic Effects of Animal Toxins," Chapter 22 in *Casarett and Doull's Toxicology*, 3rd ed., Curtis D. Klaassen, Mary O. Amdur, and John Doull, Eds., Macmillan Publishing Co., New York, 1986, pp. 706–756.

INDEX